POULTRY DISEASES: AT A GLANCE

S. Nandi
Senior Scientist
Centre for Animal Disease Research and Diagnosis
Indian Veterinary Research Institute
Izatnagar, Bareilly - 243 122, (UP), India

S. Samanta
Senior Scientist
Division of Parasitology
Indian Veterinary Research Institute
Izatnagar, Bareilly - 243 122, (UP), India

ibdc publishers

Published by

ibdc publishers

Khushnuma Complex Basement
7, Meerabai Marg (Behind Jawahar Bhawan),
Lucknow 226 001 U.P. (INDIA)
Tel. : 91-522-2209542, 2209543, 2209544, 2209545
Fax : 0522-4045308
E-Mail : ibdco@airtelmail.in

First Edition 2010

Price: Rs. 150 /-

ISBN 978-81-8189-485-4

Composed & Designed at :

Panacea Computers
3rd Floor, Agarwal Sabha Bhawan, Subhash Mohal
Sadar Cantt. Lucknow-226 002
Phone : 0522-2483312, 9335927082
E-mail : prasgupt@rediffmail.com

Printed at:

Salasar Imaging Systems
C-7/5, Lawrence Road Industrial Area
Delhi - 110 035
Tel. : 011-27185653, 9810064311

.................dedicated to the
Almighty
for everything ....

Preface

Poultry farming has become a profitable industry and has enormous potential to bring about rapid economic growth. Poultry farming in India has been transformed from backyard poultry in 1960s to a full-fledged agro-industry today at a rapid pace in just over 4 decades. India ranks fourth in egg production and fifth in broiler production in the world. The layer industry is growing at the rate of 5-7% against the broiler industry at 10-12% per annum. However, the per capita availability of eggs and broiler meat is far below the ICMR recommended level of 180 eggs and 11 Kg meat per annum. Currently, the poultry industry has the strength of 215 million layers and 1600 million broilers to produce in a year 1.9 MMT of meat and 44 billion eggs. The poultry industry is contributing about Rs 350 billion to the national GDP. Birds have been threatened by a number of infectious diseases and nutritional diseases for quite a long time. Diagnosis at an early stage of the onset of the disease has tremendous implication on the successful therapy and implementation of effective control programme. This book has been written to provide a comprehensive picture of different viral, bacterial, fungal, parasitic, chlamydial, spirochaetal, nutritional and metabolic diseases in the for of objective type questions and short notes. In this book most relevant and current informations in the form of objective type questions have been given in a concise, palatable and easily digestible fashion and it will be extremely useful for the students, teachers and researchers connected to the poultry industry. It contains the latest and in depth infromations in a very simple and easily understandable manner that can be easily and readily accepted by the student community.

With the implementation of new trimester/semester systems, the students have to face a number of quizzes, midterm and final examinations. This system evaluates the student's learning capability and teacher's teaching techniques on a regular basis. Mostly objective type and short notes are asked in the examinations. Again, in various competitive examinations namely IAS, IPS, IFS, CSIR, ICMR, ICAR, NET, JRF, SRF etc objective type questions are set to evaluate the depth of knowledge of a particular subject of the students. Further , a number of organization/institutions hold regular examination for admission in various degree programmes, award of merit scholarships and appointment on various jobs based on objective type questions.

It is therefore hoped that the book will be extremely useful for teachers as well as students and personnel of Veterinary Sciences and other disciplines of various institutions particularly associated with the poultry industry.

Lastly, the authors will appreciate receiving comments on the quality of books and errors if any for further improvement of the book.

S. Nandi
(snandi03@yahoo.com)
09412066583
S. Samanta

About the Authors

Dr Sukdeb Nandi graduated in Veterinary Science and Animal Husbandry in 1986 from Bidhan Chandra Krishi Viswavidyalaya, Mohanpur, Nadia, W.B. He did his Master Degree and Ph. D. in Veterinary Virology from Deemed University, Indian Veterinary Research Institute in 1988 and 1992 respectively. He has been associated with teaching Veterinary Virology in the Deemed University, Indian Veterinary Research Institute for last 16 years. He is also involved in imparting trainings on serological and molecular diagnosis of different Viral Diseases of animals. He was awarded with DBT sponsored National Biotechnology Associateship and carried out research on biotechnological and molecular techniques of diagnosing the diseases at Indian Institute of Chemical Biology, CSIR Institute at Jadavpur, Kolkata from 1995 to 1996. He was also deputed to United Kingdom in 1998-1999 to undergo 'Transfer of Molecular Biology Training' (TOMBIT) at Institute for Animal Health, Pirbight to work on molecular aspects of Foot and Mouth Disease Virus. He has carried out research on Rabies, FMD, Sheep Pox, Goat Pox, Bluetongue, Infectious Bovine Rhinotracheitis and Canine Parvovirus infections in canines. He has traveled different places in United Kingdom, Scotland and France. Presently, he is working as Senior Scientist at Centre for Animal Disease Research and Diagnosis, Indian Veterinary Research Institute, Izatnagar, U.P. He is also the author of 'Manual of Viral Disease Investigation' - a book very much useful for the scientists, teachers, students, diagnosticians and others who are working in this area. Other books in his credit are Veterinary Virology: at a glance, Avian influenza or Birdflu and Rabies : a killer disease.

Dr. Subhamoy Samanta, an eminent parasitologist, is presently working as Senior Scientist in the Division of Parasitology, at prestigious premier research establishment,

Indian Veterinary Research Institute, Izatnagar, UP. He did his B.V.Sc.& A.H degree from Bidhan Chandra Krishi Vishwa Vidyalaya,W.B in 1986 and completed M.V.Sc in Veterinary Parasitology from Birsa Agricultural University, Ranchi in 1988 and Ph.D from Deemed university, Indian Veterinary Research Institute in 1999. He was an excellent scholar with the honour of securing first division/class with scholarships throughout his career. He joined as Agricultural Research Service Scientist at IVRI, Izatnagar, UP in 1992. He has vast experiences in research, teaching of PG studies of Deemed University, IVRI and also engaged in various extension activities on surveillance and monitoring, diagnosis and control of parasitic diseases. He is associated with several research projects including DBT sponsored and All India Network Programme. He has worked on immunological and molecular based characterization and diagnosis of parasitic zoonosis viz. toxocariasis, hydatidosis, cryptosporidiosis, toxoplasmosis, fasciolosis, dirofilariosis, vector borne diseases and so on. His research findings are presented in various scientific forum and published in various international and national journals of repute. He is also involved in imparting training on various aspects of parasitic diseases at national and international level.

Content

Chapter 1

Avian Immune System

A. Fill in the blanks.

Q.1. ―――――――― and ―――――――― are the primary lymphoid organs in birds.

A. Thymus and bursa of Fabricius.

Q.2. In thymus, ―――――――― is densely packed with lymphocytes and ―――――――― is less densely packed with lymphocytes.

A.. Cortical area and medulla.

Q.3. First line of defense is provided by innate immune mechanisms such as ――――――――, ―――――――― and ――――――――(cells/biomolecule).

A. Phagocytic cells (heterophils, macrophages) ; NK cells ; C′.

Q.4. ――――――――, ―――――― and ―――――――― can act as antigen presenting cells.

A. Macrophages ; dendritic cells ; B cells.

Q.5. $\gamma\delta$ T cells constitute ―――――――― % of the circulating lymphocytes.

A. 30-50.

Q.6. Th1 cells produces ――――――――, ―――――――― and ―――――――― cytokines.

A. Interferon γ, TNF α and IL-2.

Q.7. Th2 cells produces ――――――――, ―――――――― and ―――――――― cytokines.

A. IL-4, IL-5 and IL-10.

Q.8. Chicken have 3 major classes of immunoglobulins namely ――――, ――― and ――.

A. IgM, IgG and IgA.

Q.9. ——————, ——————— and ——————— cytokines stimulates the class switching from IgM to IgG production.

A. IL-4, TGF-β and IFN γ.

Q.10. Avian IgG is larger than its mammalian counterpart and called ———————.

A. IgY.

B. Answer the followings.

Q.1. What are the different secondary lymphoid organs in birds?

A. Spleen, bone marrow, gland of Harder, conjunctival associated lymphoid tissue (CALT), bronchial associated lymphoid tissue (BALT) and gut associated lymphoid tissue (GALT).

C. Write true or false about the followings.

Q.1. Bursa can also act as secondary lymphoid organ.

A. True.

Q.2. Chickens have a higher proportion of γδ T cells than mouse or human.

A. True.

Q.3. Avian T cells have two surface receptors that bind antigens.

A. TCR αβ and TCR γδ.

Q.4. Th cells (CD4+ cells) recognize and process antigen in conjunction with MHC I molecules.

A. False.

Q.5. Th cells (CD8+ cells) recognize and process antigen in conjunction with MHC II molecules.

A. False.

Q.6. T, B, macrophages and dendritic cells can secrete cytokines.

A. True.

Q.7. Th1 cells largely promote the cell mediated immunity.

A. True.

Q.8. Bile is a rich reservoir of IgA in birds.

A. True.

Chapter **2**

VIRAL DISEASES

Newcastle Disease

A. Fill in the blanks.

Q.1. ND virus is under the genus ------ and family --------.

A. Avulavirus ; Paramyxoviridae.

Q.2. -------, -------- and -------- families form the order Mononegavirales.

A. Paramyxoviridae ; Filoviridae ; Rhabdoviridae.

Q.3. There are --------- serogroups of avian paramyxoviruses (APMV).

A. APMV 1 to 9. NDV is under APMV 1.

Q.4. The term Newcastle was coined by ---------.

A. Doyle.

Q.5. NDV can be grouped as --------, ------- and ---------- based on the chick embryo mortality at ---, ----- and ----- hours respectively after allantoic inoculation.

A. Velogenic ; mesogenic ; lentogenic ; < 60 h ; 60-90 h ; >90 h.

Q.6. NDV of low virulence do not form plaques in cell cultures without addition of --, ----- or ------- to the agar overlay.

A. DEAE ; Mg ++ ; trypsin.

Q.7. HN0 precursor is produced in some strains of NDV namely ------ and ----- but not in lentogenic ----- strain, mesogenic ------ strain and Velogenic ---- and ----- strains.

A. Ulster 2C; Ulster D26; Hitchner B1; Beaudette C; Australia Victoria; Italien.

Q.8. NDV has been demonstrated in at least – – – species from – – – of the 50 orders of birds.

A. 241 ; 27.

Q.9. The incubation period of ND varies from – – – to – – – –.

A. 2 to 15 days (average 5-6 days).

Q.10. – – – – – – country ban the use of any vaccine against ND in birds.

A. Sweden. But Netherlands enforce vaccination of all poultry against ND.

Q.11. – – – – – –, – – – – – and – – – – – methods are used for administration of lentogenic NDV vaccines.

A. Intranasal instillation, eye drop and beak dipping.

Q.12. – – – – – – – – and – – – – – – –. methods are used for administration of mesogenic NDV vaccines.

A. Wing web stabbing and I/M routes.

Q.13. NDV vaccines are available along with vaccines against – – –, – – – – – –, – – – and – – – – – – – – viral diseases.

A. IBV, IBDV, EDS-76 virus and Reovirus.

B. Answer the followings.

Q.1. How many genera are there under the Paramyxoviridae family and what are the different members under them?

Genera	Viruses
Rubulavirus	Mumps, mammalian PI-2 and 4.
Respirovirus	Mammalian PI-1 and 3.
Morbillivirus	Measles, CDV, RP and PPR
Avulavirus	ND virus, Avian paramyxoviurs 2-9
Henipavirus	Hendra, Nipah

Q.2. How many different forms of ND are there?

A. Doyle's form, Beach's form, Beaudette's form, Hitchner's form and asymptomatic enteric form.

Q.3. What are the different synonyms of ND?

A. Pseudofowl pest, avian pest, avian distemper, Ranikhet disease, Tetelo disease, Korean fowl plaque and avian pneumoencephalitis.

Q.4. What kind of symptoms are produced by ND virus in humans?

A. In humans, there is unilateral or bilateral reddening of the eyes, excessive lacrymation, oedema of the eyelids, conjunctivitis and subconjunctival haemorrhages.

Q.5. Why is it called Newcastle disease?

A. First outbreak of the disease occurred in 1926 in Java, Indonesia and in Newcastle upon-Tyne in England. So, it is called Newcastle disease.

Q.6. What are the different species' RBCs agglutinated by NDV?

A. NDV can agglutinate RBCs of amphibian, reptilian and avian cells besides cattle sheep, goat, swine and horse.

Q.7. What are the different avian viruses causing immunosuppression?

A. NDV, IBDV, CIAV, MDV, Reovirus etc.

Q.8. What are the different seed viruses used to produce oil adjuvanted NDV vaccines?

A. Ulster 2C, B1, La Sota, Roakin.

Write true or false about the followings.

Q.1. The enzyme neuraminidase is present in all the members of Avulavirus genus.

A. True.

Q.2. La Sota vaccine has been shown to be present in most of the reproductive organs after vaccination.

A. True.

Q.3. Lentogenic viruses do not usually cause disease in adult birds but in young susceptible birds cause respiratory symptoms.

A. True.

Q.4. Turkeys are as susceptible as chickens to infections with

ND virus but produce less clinical signs.

A. True.

Q.5. Ducks and geese are clinically resistant to even NDV most virulent for chickens.

A. True.

Q.6. Mostly IgA with some IgG is found in Harderian gland following ocular but not parenteral infection of NDV in chickens.

A. True.

Q.7. Inactivated oil adjuvanted vaccines are not as adversely affected by maternal immunity as live vaccines and can be used in day old chicks.

A. True.

Chapter 3
Infectious Bronchitis

A. Answer the followings.

Q.1. What are the clinical signs of infectious bronchitis (IB) in chickens.

A. It is characterized by coughing, sneezing and tracheal rales, gasping and nasal discharges.

Q.2. What kind of symptoms are produced by nephropathic strain of IBV in chickens?

A. There is swollen and pale kidneys with tubules and ureter distended with urates.

Q.3. What are the different diagnostic tests available to diagnose the IB?

A. AGPT, HI, FAT, IPT, ELISA, VNT and RT-PCR.

Q.4. Name the vaccines strains of IBV used for preparation of vaccines.

A. Massachusetts, Connecticut and Arkansas.

Q.5. What kind of symptoms are seen in embryos when embryonated chicken eggs are inoculated with IBV.

A. Dwarfing and curling of embryos.

B. Write true or false about the followings.

Q.1. The spikes of infectious bronchitis virus are not packed as closely as the rod shaped spikes of paramyxoviruses.

A. True.

Q.2. IBV generally infects baby chicks to produce disease.

A. True.

Q.3. IBV infected chickens have serous, catarrhal or caseous exudates in the trachea, nasal passage and sinuses.

A. True.

Q.4. ND is generally more severe in chickens than IB.

A. True.

Q.5. ILT generally produces more severe respiratory symptoms than IB but spreads more slowly.

A. True.

Q.6. In infectious coryza, facial swelling is more common but in IB, it is less common.

A. True.

Q.7. Inactivated oil adjuvanted vaccines are used at the point of laying in breeders and layers.

A. True.

Q.8. Infectious bronchitis virus is considered sensitive to most of the common disinfectants.

A. True.

Q.9. The infectious bronchitis virus can be transmitted through eggs.

A. False.

Q.10. Both live attenuated and inactivated vaccines are available against IB.

A. True.

C. Fill in the blanks.

Q.1. HI and most of the virus neutralizing antibodies are induced by ———— protein of IB virus.

A. S1.

Q.2. IBV replicates well in developing chicken embryo following inoculation by – route.

A. Allantoic.

Q.3. IBV can grow in —————, ———— and ————— primary cells.

A. Chicken embryo kidney, Chicken embryo liver and chicken kidney cells.

Q.4. Live attenuated virus vaccines against IB are used in – – – – birds and for initial vaccination of – – – – – – and – – – – – – –.

A. Broiler ; breeders ; layers.

Q.5. The spike protein of infectious bronchitis virus is called S which comprises of – – – and – – – – – – polypeptides.

A. S1 and S2.

Q.6. Infectious bronchitis virus is under genus – – – – – – – and family – – – – – – – –.

A. Coronavirus ; Coronaviridae.

Q.7. The surface projections of infectious bronchitis virus – – – – – – – shaped.

A. Club.

Q.8. Infectious bronchitis virus contains – – – – – – – – – – – – RNA as genome.

A. Positive sense single stranded.

Q.9. The best known serotypes of IB virus are –.

A. Massachusetts, Connecticut, Holland, Arkansas-99, JMK and Florida.

Q.9. Some serotypes of IB virus causing kidney disease are called – – – – – – – – – – – – – –.

A. Nephrotropic or nephropathogenic.

Q.10. – – – serotype of virus is the one most commonly included in commercial vaccines.

A. Massachusetts.

Chapter 4

Infectious Laryngotracheitis

A. Answer the followings.

Q.1. What are the characteristic symptoms of ILT in chickens?

A. It is characterized by signs of respiratory depression, gasping, coughing with bloody mucous, marked dyspnoea and high mortality.

Q.2 How many hollow capsomeres are there in an ILT virion?

A. 162.

Q.3. Which species of birds are refractory to ILT infection?

A. Starlings, sparrows, crows, doves, ducks, pigeons and guinea fowls.

Q.4. What are the tests used for diagnosis of ILT virus?

A. AGID, VNT, FAT, IPT, ELISA and PCR.

B. Fill in the blanks.

Q.1. ILT virus is under the subfamily ------ and family -------.

A. Alphaherpesvirinae and Herpesviridae. ILT virus is also known as Gallid herpesvirus1.

Q.2. ------- and ------- are preferred cell culture system for isolation of ILT virus.

A. Chicken embryo liver and chicken kidney.

Q.3. ILT virus remains in ---------- of chickens during latency.

A. Trigeminal ganglia.

Q.4. CAM of ------- days old embryonated chicken eggs are preferred to isolate the ILT virus.

A. 9-12.

Q.5. ———— titre of virus is required to induce immunity by routes other than oral route.

A. $\geq 10^2$ pfu/ml.

Q.6. ———— titre of virus is required to induce immunity by oral route.

A. 10^5 EID50.

Q.7. ILT virus is under the genus————.

A. Iltovirus.

C. Write true or false about the followings.

Q.1. Embryonated eggs of turkeys and chickens are susceptible, duck eggs are susceptible to a lesser degree and eggs of guineafowls and pigeons are not susceptible to ILT virus.

A. True.

Q.2. Meat type males are more susceptible than meat type females to ILT virus.

A. True.

Q.3. In high environmental temperature (35°C) high mortality is observed in heavy adult breeds than in light adult breeds.

A. True.

Q.4. Maternal antibodies to ILT virus is not transmitted to offspring via the eggs.

A. False. It is transmitted.

Q.5. Maternal antibodies do not confer protection against infection or interfere with vaccination.

A. True.

Q.6. I/C inclusion bodies are found in respiratory and conjunctival epithelial cells.

A. False. I/N.

Q.7. Modified live virus vaccines are used to immunize the birds against ILT.

A. True.

Q.8. Chicken can be vaccinated successfully as early as 1 day of age.

A. True.

Q.9. The ILT virus DNA replication occurs by a rolling circle mechanism with the formation of concatemers.

A. True.

Q.10. ILT virus can replicate in LMH cells, a continuous avian cell line derived from a chemically induced chicken liver tumour.

A. True.

Q.11. Although, ILT affects chickens of all ages but most characteristic signs are observed in adult birds.

A. True.

Q.12. ILT virus can spread through egg.

A. True.

Chapter 5

Avian Influenza

A. Answer the followings.

Q.1. What are the synonyms or different names of avian influenza?

A. Fowl plaque, fowl pest.

Q.2. What is the definition of OIE list A disease?

A. These are transmissible diseases that have the potential for very serious and rapid spread, irrespective of national borders which are of major importance in the international trade of animals and animal products.

Q.3. What are the two different types of mutation observed in AI viruses?

A. Antigenic shift and antigenic drift.

Q.4. What are the clinical signs produced by HPAI and LPAI in chickens?

A. The HPAI can cause serious infections in some species of birds on a farm while leaving others unaffected. The clinical signs are variable. The typical symptoms are sinusitis, lacrymation, oedema of head, cyanosis of the head, comb and wattle and green to white diarrhoea. Haemorrhagic lesions are observed on the comb and wattle of turkeys. Anorexia, coughing, sneezing, blood tinged oral and nasal discharge, ecchymoses on the shanks of feet, neurological symptoms and reduced egg production, loss of egg pigmentation and deformed or shell less eggs are also commonly observed signs. Sudden death may occur with other symptoms. Most of the flock usually dies. In ducks, the most common symptoms are sinusitis, diarrhoea and increased mortality.

The low pathogenic avian influenza (LPAI) viruses usually cause subclinical or mild illness. The symptoms may include decreased egg production, mild to moderate respiratory disease and low mortality rate (5%). More severe disease mimicking highly pathogenic avian influenza with high mortality (up to 50%) may also be seen in birds infected with LPAI when the condition is aggravated due to presence of secondary bacterial infection or other intercurrent viral diseases.

Most of the AI infections do not produce any recognizable disease in free living or wild birds. All these birds are the asymptomatic reservoirs of the AI viruses. Wild waterfowl and other aquatic birds (ducks, geese, shorebirds, gulls, swans, terns) are the reservoir of all AI viruses. Natural infections of wild birds with HPAI are rare. Viruses of H5 or H7 subtypes isolated from the free living birds are almost invariably of low pathogenicity for poultry. Seasonal coming of migratory waterfowls may allow the regular addition of new viruses to the diverse domestic poultry virus pool in the backyard poultry system.

Q.5. What are the samples to be collected from AI infected birds for diagnosis or virus isolation?

A. Tracheal or cloacal swabs from live or dead birds.

B. Write true or false about the followings.

Q.1. Cases of transfer of swine influenza viruses to humans have been reported more frequently than AI viruses.

A. True.

Q.2. Many AI viruses have been isolated from wild birds without any symptoms.

A. True.

Q.3. AI viruses from wild birds have been mostly LPAI for domestic poultry.

A. True.

Q.4. Antigenic drift in influenza viruses arise from point mutations in HA/NA gene that results in minor antigenic changes in coding proteins.

A. True.

Q.5. Antigenic shift is due to the genetic reassortment between the gene segments of two influenza viruses that infect the same cell and production of progeny viruses with new HA and NA antigens on the surface.

A. True.

Q.6. HPAI and LPAI are grouped under the OIE List A and List B diseases respectively.

A. False. HPAI (OIE List A) ; LPAI (Neither List A or List B).

Q.7. Although, H5 and H7 subtypes of AI viruses can produce HPAI, not all but a small percentage of H5 and H7 are responsible for causing HPAI rest are LPAI.

A. True.

Q.8. HPAI virus is transmitted by direct contact or indirect contact through aerosol droplets or virus contaminated fomites.

A. True.

C. Fill in the blanks.

Q.1. AI viruses bind to ––––––––––––––––– receptors.

A. N-acetylneuraminic acid alpha 2,3 galactose (NeuAcα2,3-Gal) linkages on sialyloligosaccharides.

Q.2. Human and Swine influenza viruses bind to –––––– – receptors.

A. N-acetylneuraminic acid alpha 2,6 galactose (NeuAcα2,6-Gal) linkages on sialyloligosaccharides.

Q.3. In 1997, AI virus of ––––––– subtype infected 18 people with 6 deaths in –––.

A. H5N1 ; Hong Kong.

Q.4. In 1957 and in 1968 human pandemics were caused by – –––––– and –––––– subtypes of influenza virus.

A. H2N2 ; H3N2.

Q.5. In 1957 and in 1968 human pandemics were caused by reassortment of –––––– and ––––––– genes of influenza viruses respectively.

A. HA, NA and PB1 (1957) ; HA and PB1 (1968).

Q.6. –––––––– (species of animal) acts as mixing vessel of avian and mammalian influenza viruses with the evolvement of reassortants/ new strains.

A. Swine.

Q.7. Healthy wild birds in the order ––––––– and –––––––– are asymptomatic reservoirs of AI viruses.

A. Anseriformes ; Charadriiformes.

Q.8. AI viruses are under the genus –––––––––– and family ––––––––––.

A. Influenza virus A ; Orthomyxoviridae.

Q.9. There are two surface projections in AI viruses namely ––––– and ––––––––.

A. Haemagglutinin and Neuraminidase.

Q.10. Avian influenza virus is isolated in –––––– days old embryonated chicken embryo inoculated by allantoic cavity.

A. 9-11.

Q.11. ––––and ––– cells are used for plaque assay and virus neutralization of AI virus.

A. CEF and chicken kidney.

Q.12. The HA of HPAI virus can be cleaved by –––––– proteases present in ––––––.

A. Ubiquitous furin (subtilisin) ; numerous visceral organs, nervous systems and cardiovascular system.

Q.13. The HA of LPAI is cleaved by ––––––– enzyme found in ––––––––.

A. Trypsin or trypsin like enzyme ; respiratory and intestinal system.

Q.14. AI viruses have been isolated from more than –––– species of free living birds representing –––– different orders.

A. 90 ; 13.

Q.15. AI virus is isolated in −−−−−days old embryonated chicken eggs inoculated by −−−−−−− route at the rate of −−−−−−ml /egg.

A. 10-11 ; allantoic ; 0.2 ml.

Chapter 6
Infectious Bursal Disease

A. Fill in the blanks.

Q.1. Infectious bursal disease (IBD) has a predilection of — — — — — — — — — — — — —.

A. Bursa of Fabricius.

Q.2. IBDV is under the genus — — — — — — — — and family — — — — — — — — — — —.

A. Avibirnavirus ; Birnaviridae.

Q.3. The IBDV contains — — — — — — — — segments of — — — — stranded RNA.

A. 2 ; double.

Q.4. Conformational dependent (discontinuous) epitopes are found on — — — — protein of IBDV, whereas conformational independent (continuous) epitopes are found on — — — protein.

A. VP2 ; VP3.

Q.5. Antigenic specificity determinants are present on — — — — — protein of IBDV.

A. VP2.

Q.6. Serotype — — — — — —of IBDV is pathogenic, whereas serotype — — — is non-pathogenic.

A. 1 ; 2.

Q.7. — — — — — — — route of inoculation is the better method for isolation of IBD virus in embryonated chicken egg than — — — — — — — and — — — — —routes.

A. CAM ; allantoic route ; yolk sac.

Q.8. — — — — — — — — — — — — — —cells/cell lines can be used for isolation of IBD virus.

A. CEF, RK-13, Vero and BGM-70 (baby grivet monkey cells).

Q.9. Variant strain that does not cause obvious clinical signs are found in ———————(country), whereas highly virulent strains are found in ——————————(countries).

A. U.S. ; Africa, Asia and South America.

Q.10. All breeds of chickens are susceptible but ——————— (breed) exhibits most severe reaction with high mortality when infected with IBD virus.

A. WLH.

Q.11. The period of greatest susceptibility to IBD virus infections in chickens is — to— weeks of age.

A. 3 to 6.

Q.12. Bursectomised chickens challenged with virulent virus produced —————— times less IBD virus than control birds.

A. 1000.

Q.13. IBD virus infections in chickens causes————% mortality and ————% morbidity.

A. 20-30 ; 100.

Q.14. The common group antigens for both serotypes of IBD virus are on ——— and ——— proteins.

A. VP2 and VP3.

Q.15. ——————— protein has serotype specific antigens that induce virus neutralizing antibodies.

A. VP2.

Q.16. The half life of maternal antibodies to IBD virus is between ——— to ——— days.

A. 3 to 5.

Q.17. Suppression of the antibody response to NDV was greatest in chicks infected with IBD virus at — day of age, moderate at - days of age and negligible at ——— days of age.

A. 1 ; 7; 14-21.

Q.18. IBD virus infected birds can be susceptible to ————, ——————— and —————— diseases.

A. Inclusion body hepatitis, coccidiosis, MD, ILT, IB, CIA, Salmonella and Colibacillosis.

Q.19. The strains of IBD virus used in vaccine preparation are ————, ———, ———, —— and ——————.

A. Mild, mild intermediate, intermediate, intermediate + and hot.

Q.20. Highly virulent (hot), intermediate and avirulent strains break through maternal virus neutralizing antibody titers of ———————, ———————— and —————————.

A. 1 :500 ; 1 :250 and less than 1 :100.

B. Answer the followings.

Q.1. What is the synonyms of IBD?

A. Gumboro disease.

Q.2. Why is it called gumboro disease?

A. Gumboro is the name of the place where first outbreak of the disease was reported.

Q.3. What are the different structural proteins present in the IBD virus?

A. Five structural protein namely VP1-VP5. VP2 and VP3 are major and VP1 and VP4 are minor proteins.

Q.4. How many serotypes of IBD virus are there?

A. Two. Serotype 1 and serotype 2.

Q.5. What are the two different strains of serotype 1 IBDV?

A. Classic/standard and Variant.

Q.6. What are the clinical signs of IBD virus infections in chickens?

A. The incubation period is 2-3 days. Birds have soiled vent feathers, whitish diarrhoea, anorexia, depression, ruffled feathers, trembling, severe prostration and finally death occur.

Q.7. Describe the gross pathology in chickens infected with IBD virus.

A. Infected birds are dehydrated with darkened

discolouration of pectoral muscle and thigh muscle, mucous in intestine with renal changes. On day 3 bursa begins to increase in size and weight. On day 4 bursa is double the normal weight, size begins to recede. On day 5 bursa is normal in weight but atrophied. On day 8 bursa is 1/3 of the normal weight.

C. Write true or false about the followings.

Q.1. Extreme kidney function occurs in IBD infections in birds.

A. True.

Q.2. Two serotypes of IBDV can be separated by FAT and ELISA.

A. False.

Q.3. Two serotypes of IBDV can be separated by SNT.

A. True.

Q.4. Immunization with serotypes 2 of IBDV gives protection in birds against serotype 1.

A. False.

Q.5. Chickens and turkeys are the natural hosts of the IBD virus.

A. True.

Q.6. Guinea fowls inoculated with the IBD virus do not develop lesions or antibodies.

A. True.

Q.7. Both light and heavy breeds of chickens are equally susceptible to IBD virus.

A. True.

Q.8. 3 weeks old chickens treated with cyclophosphamide are refractory to IBD virus challenge.

A. True.

Q.9. Birds with surgically removed bursa at 4 weeks of age are refractory to IBD virus challenge.

A. True.

Q.10. IBD virus is transmitted through eggs.

A. False.

Q.11. The carrier state is reported in chickens infected with IBD virus.

A. False.

Q.12. Very virulent IBD virus strains cause a greater decrease in thymic weight index, severe lesions in caecal tonsils, thymus, spleen and bone marrow.

A. True.

Q.13. Both live attenuated and killed vaccines are available for IBD.

A. True.

Q.14. Inactivated and live attenuated virus vaccines made from variant strains protected chickens from disease caused either variant or standard strains.

A. True.

Q.15. Adult birds are resistant to oral exposure to the IBD virus.

A. True.

Q.16. Instead of getting infected with disease, adult birds produce antibody after I/M or S/C inoculation of IBD virus.

A. True.

Q.17. IBD virus can suppress the antibody response to other vaccines.

A. True.

Q.18. Chickens infected with IBD virus at 1 day of age were completely deficient of serum IgG and produce only monomeric IgM.

A. True.

Q.19. The number of B cells in peripheral blood is decreased following infection with IBD virus, but T cells were not appreciably affected.

A. True.

Q.20. IBD virus affects immature or precursor B cells to a great extent than mature B cells.

A. True.

Q.21. CAM of 9-11 days old embryo is the most sensitive route for isolation of the IBD virus.

A. True.

Q.22. Variant strains of IBD virus differ from standard strain, is that they induce splenomegaly, liver necrosis but little mortality.

A. True.

Q.23. Virus neutralization test is the only serological test for differentiation of isolates of IBD virus with antigenic variation.

A. True.

Q.24. Intermediate strains of IBD virus vary in their virulence and can induce bursal atrophy in day old and 3 weeks old SPF chickens.

A. True.

Q.25. If maternal antibody titer is < 1:1000, chickens may be vaccinated with avirulent strains of IBD virus.

A. True.

Q.26. The IBD vaccine virus replicates in the thymus, spleen and cloacal bursa where it persists for 2 weeks.

A. True.

Q.27. Inactivated oil adjuvanted vaccine is more effective in chickens that have been primed with live virus vaccine or field exposure of the virus.

A. True.

Chapter 7

Chicken Infectious Anaemia

A. Answer the followings.

Q.1. Write down the synonyms of the chicken infectious anaemia (CIA).

A. CIA is known as haemorrhagic syndrome, anaemia dermatitis or blue wing disease. It is characterized by aplastic anaemia, generalized lymphoid atrophy with immunosuppression.

Q.2. What is the genus and family of the CIA virus?

A. Gyrovirus ; Circoviridae.

Q.3. Describe the genome of the CIA virus.

A. The genome of CIA virus consists of single stranded circular covalently linked DNA of negative sense.

Q.4. What is the site of replication of CIA virus in chickens?

A. The replication of CIA virus in chickens occurs mainly in haemocytoblasts in the bone marrow and T cell precursors in the cortex of the thymus.

Q.5. How the CIA virus can be inactivated?

A. 1% glutaraldehyde for 10 min at room temperature, 0.4% BPL for 24 hours at 4°C or 5% HCHO for 24 at room temperature.

Q.6. Describe the clinical signs of CIA infections in chickens.

A. Affected birds are anaemic, depressed and pale with reduced body weight. Morality is less than -30%. Blood becomes watery and clotting time is increased. There is decreased number of RBCs, WBCs and thrombocytes.

Q.7. What are the different immunosuppressive avian viruses?

A. MD, CIA, REV, IBD and Reovirus.

Q.8. What are the gross lesions found in CIA infected chickens?

A. Thymic atrophy with almost absence of thymic lobes, thymic remnants become dark red colour, bone marrow atrophy with yellowish or pink in colour. Bursal atrophy is less commonly associated with CIA virus infection.

Q.9. What are the different materials to be used for the isolation of CIA virus?

A. Most of the tissues, buffy coats, livers and lymphocytes from spleen.

B. Fill in the blanks.

Q.1. CIA virus was first isolated in the year ---- in ----(country) by -------.

A. 1979 ; Japan ; Yuasa and co-workers.

Q.2. Type II particles of CIA virus look like ---------- structure under EM.

A. Cog wheel.

Q.3. VP3 associated with nuclei of CIA virus infected cell is called -------.

A. Apoptin.

Q.4. ------ T cell lines and -------- B cell lines are suitable for propagation of CIA virus.

A. T cell lines (MDCC-MSB1 and MDCC-JP2) ; B cell lines (LSCC-1104B1).

Q.5. ------ cell line is the best cell line for propagation of CIA virus.

A. MDCC-CU147.

Q.6. Antibodies to CIA virus have been detected in ----- (species of bird) but not in -----------, -------- and ------- (species of birds).

A. Japanese quail ; turkey, pigeons, crow and ducks.

Q.7. ---- chicks seemed to be more susceptible to disease than --- and ---chicks.

A. S13 (MHC :B13B13) ; N2a (MHC :B21B21) and P2a (MHC :B19B19).

Q.8. Maternal antibodies give protection against CIA virus infections for – – – – weeks.

A. 3.

C. Write true or false about the followings.

Q.1. A 50 Kda viral protein (VP1) is the only protein detected in CIA virus.

A. True.

Q.2. VPQ protein of CIA virus infected cell is a nonstructural protein.

A. True.

Q.3. Neutralizing antibodies were induced by VP1 and VP2 protein not either VP1 or VP2.

A. True.

Q.4. Commercial disinfectants based on soap or orthodichlorobenzene are not effective against CIA virus.

A. True.

Q.5. CIA virus is resistant to treatment with 90% acetone.

A. True.

Q.6. CIA virus is resistant to heating 56ºC or 70ºC for 1 hour and 80ºC for 15 minutes.

A. True.

Q.7. CIA virus is resistant to pH 3.0 for 3 hours.

A. True.

Q.8. There is only one serotype of CIA virus with no antigenic difference.

A. True.

Q.9. CIA virus can be propagated in cell culture, 1 day old chick and chicken embryo.

A. True.

Q.10. Chicken is the only known host for CIA virus.

A. True.

Q.11. All ages are susceptible but susceptibility to diseases decreases in immunologically intact chicks during the first

1-3 weeks of life.

A. True.

Q.12. Both horizontal and vertical transmission have been reported in CIA virus.

A. True.

Q.13. Vertical transmission of CIA virus occurs through infected semen and infected chick is born.

A. True.

Q.14. CIA virus can not establish a latent infection in chickens.

A. False.

Q.15. Outbreaks of chicken infectious anaemia in field flocks are mostly associated with so called haemorrhagic syndrome.

A. True.

Q.16. CIA virus damages both haemotopoietic and lymphopoietic tissues.

A. True.

Q.17. Both inactivated and live attenuated vaccines are available.

A. True.

Q.18. The CIA virus replicates by rolling circle model.

A. True.

Q.19. Vertical transmission through hatching eggs is most important mode of dissemination of CIA virus.

A. True.

Q.20. Thymic atrophy is the most consistent lesions and bone marrow atrophy is the most characteristic lesions seen in affected chickens.

A. True.

Q.21. Liver is the important source of CIA virus for diagnosis because it contains high concentration of virus.

A. True.

Q.22. ELISA is more sensitive than VNT and direct FAT in diagnosis of CIA infection in chickens.

A. True.

Chapter **8**

Egg Drop Syndrome - 76

A. Answer the followings.

Q.1. What is the genus and family of EDS-76 vrius?

A. Atadenovirus ; Adenoviridae.

Q.2. Write down the haemaggulutination properties of EDS virus.

A. EDS virus haemagglutinates the RBCs of chicken, duck, turkey, geese, pigeon and peacock but does not haemagglutinate the RBCs of rat, rabbit, horse, sheep, cattle, goat and pig.

B. Fill in the blanks.

Q.1. The EDS virus has — — — penton fiber, whereas subgroup I viruses have — — — —.

A. 1 ; 2.

Q.2. — — — — — — — destroys the virus receptor on chicken RBCs, whereas — — — and — — — — — has no effect.

A. α trypsin ; trypsin and neuraminidase.

Q.3. EDS virus replicates in — — — — — —, — — — — — — — and — — — — — — — cell culture.

A. Duck kidney, duck embryo liver and duck embryo fibroblast cell culture.

C. Write true or false about the followings.

Q.1. EDS virus is the sole member of subgroup III avian adenovirus and unrelated serologically to subgroup I and II.

A. True.

Q.2. Only one serotype of EDS-76 virus is reported.

A. True.

Q.3. EDS is characterized by thin shelled or shellless eggs and major cause of reduced or lost egg production throughout the world.

A. True.

Q.4. It is transmitted vertically through eggs but not horizontally.

A. True.

Q.5. EDS-76 virus contains high A+T content.

A. True.

Q.6. HA of EDS virus is resistant to heating at 56°C but destroyed at 70°C for 30 min.

A. True.

Q.7. Only one serotype and three genotypes of EDS are reported.

A. True. Three genotypes are U.K., Europe and Australia.

Q.8. EDS virus grows well in allantoic sac of embryonated duck or goose eggs but not in embryonated chicken eggs.

A. True.

Q.9. EDS virus affects only laying hens and natural hosts for EDS are ducks and geese.

A. True.

Q.10. Quail (*Coturnix coturnix japonica*) have been susceptible to EDS virus and develop signs of EDS.

A. True.

Q.11. Broiler breeders and heavy breeds producing brown eggs are more severely affected than white egg producers.

A. True. 3 times more.

Q.12. Inactive ovary and atrophied oviducts are observed in naturally occurring outbreaks of EDS.

A. True.

Q.13. Unlike subgroups I and II, EDS virus does not appear to replicate in the intestinal mucosa.

A. True.

Q.14. EDS virus causes the latency in infected birds.

A. True.

Q.15. Chicken embryo liver cells are more sensitive than chicken kidney cells but chicken embryo fibroblast and embryonated chicken eggs are not suitable or insensitive.

A. True.

Q.16. Duck or goose cells and embryonated duck and goose eggs are also sensitive to EDS virus.

A. True.

Q.17. Oil adjuvanted vaccines are widely used against EDS.

A. True.

Chapter 9
Fowl Pox

A. Answer the followings.

Q.1. What are the synonyms of fowl pox?

A. Contagious epithelioma, avian diphtheria, Variola gallinarum.

Q.2. What are the two form of the fowl pox disease?

A. There are two forms of the disease namely cutaneous and diphtheritic. Cutaneous form is characterized by discrete nodular proliferative skin lesions on non-feathered part of the body. Diphtheritic form is characterized by fibrino-necrotic and proliferative lesions on the mucous membrane of the upper respiratory tract, mouth and esophagus.

Q.3. What are Borrel bodies?

A. Elementary bodies found in FP virus infected cells can be detected in smears prepared from lesions stained with Wright's stain or by the Gimenez method are called Borrel bodies.

B. Fill in the blanks.

Q.1. Out of 9000 bird species, – – –(nos) in – – – –(nos) orders are naturally infected with pox virus.

A. 232 ; 23.

Q.2. Fowl pox is under the genus – – – – – – and family – – – – – – – – –.

A. Avipox virus ; Poxviridae.

Q.3. – – – – – – days old embryonated chicken eggs are generally used for isolation of FP virus inoculated by – – – – route.

A. 10-12 ; CAM.

Q.4. FP virus can be grown in ————, ———————, ——— —— and ————————.

A. CEF, chicken embryo dermis and chicken kidney and duck embryonated cells.

Q.5. ——————— and ——————— cell line can support the growth of some avian pox virus.

A. QT35 of Japansese quail origin ; chicken liver cell line LMH.

Q.6. ——————— type of inclusion bodies are found in FP virus infected cells.

A. Eosinophilic A type cytoplasmic inclusion bodies (Bollinger bodies).

C. Write true or false about the followings.

Q.1. FP virus contains single copy of linear ds DNA with hairpin loop at each end.

A. True.

Q.2. HA activity is not shown by most of the FP virus.

A. True.

Q.3. Canaries are highly susceptible to canarypox virus but show resistance to turkey, fowl and pigeon pox virus.

A. True.

Q.4. Pigeon pox is very pathogenic for pigeons but causes a mild disease in turkey and chickens.

A. True.

Q.5. Ducks are susceptible to turkey pox virus but not FP virus.

A. True.

Q.6. All the strains of FP virus produce plaques in cell lines.

A. False.

Q.7. Diphtheritic form of FP and ILT produce similar signs and tracheal lesions.

A. True.

Q.8. Lesions caused by pantothenic acid or biotin deficiency in young chicks or by T-2 toxins can be confused with pox lesions.

A. True.

Q.9. FP virus encodes DNA dependent RNA polymerase.

A. True.

Q.10. Canarypox virus vaccine expressing rabies virus G is currently available commercially for use in cats.

A. True.

Chapter 10

Avian Encephalomyelitis

A. Answer the followings.

Q.1. What are the synonyms of avian encephalomyelitis.

A. Epidemic tremor and New England disease.

B. Fill in the blanks.

Q.1. AE virus is under the genus ------ and family -----------.

A. Enterovirus ; Picornaviridae.

Q.2. AE virus has ---- symmetry with ------- capsomeres.

A. icosahedral ; 60.

Q.3. There are two pathotypes of AE virus ---- and ---------.

A. Field strain (enterotropic) and embryo adapted strain(neurotropic) or VR strain.

Q.4. ------ route of --------- days old embryonated chicken eggs are generally used for isolation of virus.

A. Yolk sac ; 5-7.

Q.5. ------, -------, ------- and ------- cells from young chickens are used to cultivate both the strains of AE virus.

A. CEF, chicken embryo kidney, chicken embryo neurological cells and pancreatic cells.

Q.6. AE virus causes -------% morbidity and -----% mortality.

A. 40-60 ; 25.

C. Write true or false about the followings.

Q.1. I/C crystalline array observed in Purkinje cells from the brains of AE virus infected chickens.

A. True.

Q.2. AE virus highly resistant to $CHCl_3$, acid, trypsin and pepsin and DNase but susceptible to HCHO fumigation.

A. True.

Q.3. Field strain of AE virus are nonpathogenic, whereas neurotropic strains cause severe neurological signs.

A. True.

Q.4. Both strains of AE virus can replicate in embryos but natural/field strains do not cause obvious signs or lesions.

A. True.

Q.5. Both horizontal and vertical transmission occur in AE virus.

A. True.

Q.6. Carrier state has been reported in AE virus infected birds.

A. False.

Chapter 11
Reovirus Infections

A. Answer the followings.

Q.1. What are the different clinical signs produced by avian reovirus in poultry?

A. It causes arthritis/tenosynovitis, stunting syndromes, respiratory disease, enteric disease, immunosuppresion and malabsorption syndrome. Reovirus also causes growth retardation, pericarditis and myocarditis.

Q.2. Write down about the genome of reoviruses?

A. Reoviruses have 10 double stranded RNA segments which are categorized as large, medium and small.

Q.3. What are the different cells used for isolation of reovirus?

A. Primary cell culture of embryo, chicken lung, kidney, liver, macrophage and testicle are used for growth of reovirus.

Q.4. What are the different cell lines support the growth of revovirus?

A. Vero, BHK-21, CRFK, Georgia bovine kidney (GBK), rabbit kidney, porcine kidney, Japanese quail cell line (QT35).

B. Fill in the blanks.

Q.1. Avian reovirus is under the genus ––––– and family–––––.

A. Orthoreovirus ; Reoviridae.

Q.2. The strain used for preparation of the vaccine against reovirus is –––––.

A. S1133.

Q.3. ––––– % of reoviruses isolated from chickens are apathogenic.

A. 80.

Q.4. Reoviruses replicate in ––––––– (location of the cell), have –––––– symmetry and without envelope.

A. Cytoplasm ; icosahedral.

Q.5. Reoviruses can be inactivated by ––– and –––––––.

A. 70% ethanol ; 5% HCHO.

Q.6. –––––– and ––––––– routes of ECE are used for isolation of reovirus.

A. Yolk sac and CAM.

Q.7. Reovirus causes ––––––% morbidity and ––––––– % mortality.

A. 100 ; <6.

C. Write true or false about the follwings.

Q.1. Both live attenuated and inactivated vaccines are available against avian reovirus infections.

A. True.

Q.2. Reoviral arthritis is observed in meat type chickens and can be found in lighter breeds and turkey.

A. True.

Q.3. Reoviruses are found commonly in gastrointestinal and respiratory tracts of clinically normal chickens and turkeys.

A. True.

Q.4. Reoviruses are sensitive to ether but are slightly sensitive to $CHCl_3$.

A. True.

Q.5. The vaccine strain of revovirus (S1133) can be inactivated by trypsin.

A. True.

Q.6. Reovirus infected embryos are dwarf with enlargement of liver and spleen having necrotic foci.

A. True.

Q.7. Both I/C acidophilic or basophilic inclusion bodies are found in cell lines.

A. True.

Q.8. Chickens and turkeys are natural or experimental hosts for reovirus induced arthritis.

A. True.

Q.9. Turkeys are shown to be more resistant than chickens to revovirus induced tenosynovitis.

A. True.

Q.10. Reovirus may persists for long period in the caecal tonsils and hock joints of birds.

A. True.

Q.11. Both horizontal and vertical transmission have been reported in reovirus infection in birds.

A. True.

Q.12. Reovirus interferes with HVT derived MD vaccines.

A. True.

Chapter 12
Leechi disease

A. Answer the followings.

Q.1. What are the synonyms of the leechi disease?

A. Hydropericardium syndrome, hydropericardium hepatitis syndrome, Angara disease.

Q.2. Why is the disease called leechi disease?

A. Due to accumulation of excessive fluid under the pericardium of heart, the heart looks like a peeled leechi fruit. That is why it is called leechi disease.

Q.3. What are the different diagnostic tests available for leechi disease?

A. AGPT, CIE and ELISA.

B. Fill in the blanks.

Q.1. The leechi disease was first recognized in the year ––––in –––––– (country).

A. 1987 ; Angara Goth, near Karachi, Pakistan.

Q.2. Leechi disease occurs in ––––––– weeks old broilers.

A. 3-5.

C. Write true or false about the followings.

Q.1. The virus causing leechi disease can be transmitted horizontally and vertically.

A. True.

Q.2. Latent state is reported in birds infected with virus causing leechi disease.

A. True.

Chapter 13

Duck Plaque

A. Answer the followings.

Q.1. What is the another name of duck plaque?

A. Duck viral enteritis.

Q.2. What are the different species of birds affected by duck plaque virus?

A. Duck, geese and swans.

Q.3. Write down the clinical signs of duck plaque in ducks?

A. Affected birds exhibit photophobia, half closed eyelids, inappetance, extreme thirst, droopiness, ataxia, ruffled feather, nasal discharges, soiled vents and watery diarrhoea, unable to stand, tremors of head, neck and body.

Q.4. Write down the pathological lesions of duck plaque in ducks?

A. The duck plaque is characterized by vascular damage, tissue haemorrhage, eruptions of mucosa of gastrointestinal tract, lesions of lymphoid organs and degenerative parenchymatous organs.

B. Fill in the blanks.

Q.1. Duck plaque virus is under the subfamily ——————— and family ———————.

A. α herpesvirinae ; Herpesviridae. It is also called anatid herpesvirus 1.

Q.2. DP virus naturally infects the birds under family ————— and order ——————.

A. Anatidae (ducks, geese, swans) ; Anseriformes.

Q.3. ———————— is the site of latency for DP virus in birds.

A. Trigeminal ganglion.

Q.4. DP virus can be isolated by inoculating – – – – days old EDE by – –route.

A. 9-14 ; CAM.

Q.5. – – – – – – – – – – cells/cell lines are used for isolation of DP virus.

A. Pekin or Muscovy duck embryo fibroblast, liver, kidney.

C. Write true or false about the followings.

Q.1. DP virus does not cause any haemaggulutination or haemadsorption.

A. True.

Q.2. Birds of the order Charadriiformes [herring gulls (*Larus argentatus*) and black headed gulls (*L. Ridibundus*)] are not susceptible to experimental infection with DP virus.

A. True.

Q.3. Carrier state has been reported in DP infected birds.

A. True.

Q.4. Both live attenuated and inactivated vaccines are available for DP.

A. True.

Chapter 14
Marek's Disease

A. Answer the followings.

Q.1. What is Marek's disease?

A. It is a lymphoproliferative disease of chickens characterized by infiltration of mononuclear cells in peripheral nerves and various other organs, iris and skin.

Q.2. How many serotypes of Marek's disease are there?

A. 3. Serotype 1, Serotype 2 (nononcogenic chicken isolate) and serotype 3 (herpes virus of turkey).

Q.3. How many pathotypes of MD serotype 1 are there?

A. 4. Mild, virulent, very virulent and very virulent +.

Q.4. Three different types of virus cell interactions namely ————, ———— and ———— are reported.

A. Productive, latent and transforming.

Q.5. What is growth character and plaque morphology of different serotypes of MD virus?

A.

Serotype	Cell line	Growth	Plaque morphology
Low passage serotype 1	Grows best in DEF or chicken kidney cell culture	Grows slowly	Small size plaque
Serotype 2	Grows best in CEF	Grows rapidly	Medium size plaque with large syncytia
HVT	Grows best in CEF	Grows rapidly	Large plaque

Q.6. What is the characteristic clinical feature of MD in a bird?

A. The characteristic clinical feature of MD in a bird is one leg stretch forward and other backward as a result of unilateral paresis or paralysis.

Q.7. What are the clinical signs of MD in chickens?

A. There are two forms of the disease : classical and acute. In classical form, chickens display varying degree of ataxia and flaccid paralysis of the neck or limbs beginning 8-12 days after exposure to the virus by inoculation or contact. Signs last for 1-2 days followed by rapid and complete recovery. Although, recovered chickens succumbed few weeks later with MD lymphoma. In the acute form, death occurs 24-72 hours following the onset of paralytic signs.

Q.8. What are the pathological lesions found in MD infected chickens?

A. Enlargement of peripheral nerves, loss of cross striation, gray or yellow discolouration and oedema of peripheral nerves are noticed. Lymphoma may occur in gonad (specially ovary), lungs, heart, mesentery, kidney, liver, spleen, bursa, thymus, adrenal gland, pancreas, proventriculus, iris, intestine, skeletal muscle and skin. Visceral lymphomas are common in more virulent forms of the disease. Visceral tumours can occur in the absence of gross nerve lesions. MD lymphomas in viscera are white or grayish in colour. Multiple nodular tumours in the myocardium is noticed. Erythematous involvement of the shank tegument in broiler chickens infected with virulent virus is observed and commonly known as Alabama red leg. Loss of pigmentation in the iris is known as 'Gray eye', and conjunctivitis with multifocal haemorrhages and corneal oedema is also seen.

Q.9. What are the different vaccine strains used in preparation of MD vaccines?

A. Attenuated serotype 1 strains : CVI -988 (Rispens), R2/23 ; Serotype 2 : 301B/1, SB-1 and serotype 3 : FC-126.

B. Write true or false about the followings.

Q.1. MD virus vaccine is the first effective vaccine against cancer in any species.

A. True.

Q.2. Chickens vaccinated with 3 serotypes can be superinfected with virulent MDV.

A. True.

Q.3. MD virus latency is mostly associated with CD4+ T cells, although CD8+ T cells and B cells can also be latently infected.

A. True.

Q.4. The presence of latent MD virus infections can be detected by PCR.

A. True.

Q.5. CIA virus may influence latency in MD by influencing cytokine production and Tc cells.

A. True.

Q.6. Transformation occur only in cells infected with serotype 1 of MD virus.

A. True.

Q.7. MATSA (Marek's disease tumour associated surface antigen) has been detected on cells from MD lymphomas and lymphoblastoid cell lines.

A. True.

Q.8. MATSA has been detected on lymphocytes from chickens vaccinated with HVT or nononcogenic strains of MD virus.

A. True.

Q.9. All the 3 serotypes of MD virus are oncogenic.

A. False. Only MD virus serotype 1 is oncogenic.

Q.10. Latent infection has also been reported in serotype 2 and 3 of MD virus.

A. True.

Q.11. Dander, litter and feathers from infected chickens are infectious and contain cell free MD virus.

A. True.

Q.12. Virulence or oncogenicity is associated with only serotype 1 of MD virus.

A. True.

Q.13. Serotype 1 but not serotype 2 of MDV can be grown in chicken splenic lymphocytes.

A. True.

Q.14. All the cell lines established from lymphomas of birds infected with MD display a chromosomal aberration.

A. False. Most but not all.

Q.15. Of the different syndromes produced by MD, lymphoproliferative syndromes are most frequently associated with MD.

A. True.

Q.16. MD exists in poultry producing countries throughout the world.

A. True.

Q.17. Chickens are the most important natural host for MD.

A. True.

Q.18. Quail, turkeys and pheasants are also susceptible to MD virus infections and disease.

A. True.

Q.19. Virtually all chickens including game fowl, native breeds and jungle fowls are susceptible to MDV infections.

A. True.

Q.20. In Japanese quail, natural outbreaks of MD is common.

A. True.

Q.21. Transmission of MDV from chickens to quail and quail to chickens has not been reported.

A. False.

Q.22. Natural outbreaks of MD in turkey are rare.

A. True.

Q.23. Vaccination with HVT provide no protection against MD virus challenge in turkey.

A. True.

Q.24. Both B and T cell lines have been developed from turkey lymphomas induced by MD virus inoculation.

A. True.

Q.25. Pheasants and black Francolin are susceptible to MD virus.

A. True.

Q.26. Unlike quail and turkey, pheasants are more susceptible to neurological lesions of MD.

A. True.

Q.27. MD is transmitted by direct and indirect contact apparently by aerosol route.

A. True.

Q.28. Epithelial cells in the keratinizing layer of the feather follicle contain fully infectious virus and source of virus for contamination of environment.

A. True.

Q.29. MD virus associated with feathers and danders are infectious.

A. True.

Q.30. MDV remains infectious for several months at 20-25°C and for years at 4°C.

A. True.

Q.31. Once infected, chickens appear to shed MD virus indefinitely.

A. True.

Q.32. Vertical transmission of MDV occurs in chickens.

A. True.

Q.33. Atrophy of thymus and bursa of Fabricius has not been reported in MD infected birds.

A. False.

Q.34. Atrophy of bursa of Fabricius and thymus is seen in MD.

A. True.

Q.35. The replication of MD virus occurs in feather follicle epithelium of genetically resistant and susceptible birds.

A. True.

Q.36. Only T cell MD tumour has been reported in experimentally infected turkeys.

A. False. Both T and B cells MD tumours have been reported.

Q.37. MD transformed cells mostly but not entirely nonproductive in *Vivo* and *Vitro*.

A. True.

Q.38. Both genetic strain of bird and virus strains determine the distribution as well as incidence of lymphomas.

A. True.

Q.39. Vaccines against MD does not prevent superinfection with wild type virus but prevent tumour development.

A. True.

Q.40. MD virus neutralizing antibodies are important only when cell free virus infects chickens or MD virus protein are expressed on the cell surface.

A. True.

Q.41. HVT, attenuated serotype 1 MDV and serotype 2 MDV protect the birds against early replication of virulent virus in the lymphoid organs of challenged birds and reduce latent infection.

A. True.

Q.42. CMI only is suppressed by MDV infection.

A. False. Both humoral and cell mediated immunity are suppressed.

Q.43. Chicken kidney cells and duck embryo fibroblast culture are preferred for isolation of MDV serotype 1.

A. True.

Q.44. Both serotype 1 and 2 vaccines are available as cell free and cell associated products against MD.

A. False. Only cell associated.

Q.45. HVT vaccine in genetically resistant chickens produces better immunity than HVT and SB-1 in a susceptible chicken.

A. True.

Q.46. Vaccines of serotype 2 and 3 combined give excellent protection against MD.

A. True. It is due to the synergistic effect of the two serotypes.

Q.47. In case of MD vaccines, I/M route is better than S/C route.

A. True.

Q.48. *In ovo* vaccination against MD has been reported.

A. True.

Q.49. MDV serotype 2 and 3 vaccines and R2/23 vaccines are clearly more effective than HVT.

A. True.

Q.50. Serotype 1 MD vaccines provided best protection in birds with certain B haplotypes.

A. True.

Q.51. Serotype 2 MD vaccines provided best protection in birds with certain B^5 haplotypes.

A. True.

Q.52. HVT causes clinical diseases in immunosuppressed chickens.

A. False

Q.53. SB-1 (serotype 2) suppresses humoral immunity and is immunosuppressive.

A. False. Do not suppress humoral immunity and not immunosuppressive.

C. Fill in the blanks.

Q.1. MD virus in under the subfamily ——————— and family ————————.

A. Alphaherpesvirinae ; Herpesviridae.

Q.2. ———————— is the cellular receptor of MD virus.

A. Heparan sulphate.

Q.3. Reactivation of latently infected cells can be induced by ————, ———— and ————.

A. Stress condition ; Betamethazone ; cyclosporine.

Q.4. More number of infectious virus particles are produced by ––––– serotype from cells than that of ––––– serotypes.

A. HVT ; Serotypes 1 and 2.

Q.5. –––––– and –––––– cells are used for isolation of serotypes 1, 2 and 3 of MD virus.

A. Duck embryo fibroblast (DEF) and chicken kidney.

Q.6. ––––––– inclusion bodies are seen in MD virus infected cell.

A. Type A intranuclear.

Q.7. –––––– and ––––––cell lines can be used to propagate 3 serotypes of MD virus.

A. OU 2 ; DF-1.

Q.8. Virus pocks develop on CAM of chicken embryo after inoculation with cellular MDV preparation by –––––– route.

A. Yolk sac.

Q.9. –––––––––––– (species of birds) are resistant to MD virus infections.

A. Ducks, sparrows, patridges, pigeons and peafowl.

Q.10. MDV excretion starts –––weeks after infection and maximum shedding occurs between ––––weeks.

A. 2 ; 3-5.

Q.11. Latent infection of MD has been observed in ––––––– and ––––––– cells in spinal ganglia.

A. Schwann and satellite.

Q.12. MD virus is transferred to feather follicle epithelium (FFE) of infected birds by –––– cells.

A. Infected lymphocytes.

Q.13. MD lymphoma contains –––––, –––––, –––––– and –––––––– cells.

A. Neoplastic cells, inflammatory cells, T and B cells.

Q.14. ––––––– cell is preferred to isolate the MDV 2 and 3 and attenuated MDV serotype 1.

A. CEF.

Q.15. HVT strain –––– is used for preparation of vaccine against MD.

A. FC126.

Q.16. ––––––, –––––––– and –––––––– serotypes are used for vaccine preparation against MD.

A. Attenuated MD virus serotype 1, serotype 2 and HVT.

Q.17. Between cell free and cell associated HVT vaccines, –– –––––– is more effective in the presence of maternal antibodies.

A. Cell associated.

Q.18. At least ––––––– pfu/dose of MD vaccine is required to provide better protection in layers.

A. 5000.

Q.19. In broiler ––––––pfu/dose of MD vaccine is required to provide better protection against MD.

A. ½ to 1/3 rd of the layer dose (Layer : 5000 pfu/dose).

Q.20. ––––, –––––––, –––––, and ––––– viruses are reported to have interfere with the induction of vaccinal immunity.

A. IBD, Reticuloendotheliosis virus, Reovirus and chicken infectious anaemia.

Q.21. Most effective vaccine against MD is ––––––––.

A. Original CVI 988 based on serotype II.

Q.22. MD virus is under the genus ––––––––.

A. Mardivirus.

Chapter **15**

Leukosis/Sarcoma

A. Write true or false about the followings.

Q.1. Both endo and exogenous ALV can be detected in the albumin of eggs by RT-PCR.

A. True.

Q.2. Retroviruses are icosahedral in shape.

A. True.

Q.3. Females are more susceptible to lymphoid leucosis than males.

A. True.

Q.4 In the albumin of not all eggs that have ALV give rise to infected embryos or chicks.

A. True.

Q.5. Marked enlargement of the liver occurs in Lymphoid leucosis.

A. True.

Q.6. Nodular growth occurs in bursa of Fabricius of all cases of lymphoid leucosis.

A. True.

Q.7. In lymphoid leucosis, tumours are found in liver, spleen and bursa of Fabricius.

A. True.

Q.8. No vaccine is available against Lymphoid leucosis.

A. True.

B. Fill in the blanks.

Q.1. Avian leucosis/Rous sarcoma virus is under the genus ––––– and family ––––––.

A. α retrovirus ; Retroviridae.

Q.2. Avian leucosis virus has ——————————— genes from 5′ to 3′.

A. 5′ gag-pol-env 3′.

Q.3. RSV carries ——— oncogene which code a ————kD protein having protein kinase activity.

A. src ; 60.

C. Answer the followings.

Q.1. What are the different types of tumours found in lymphoid leucosis?

A. Tumours appear as nodular, granular or diffuse or a combination of these forms. In the nodular form, lymphoid tumours vary 0.5 mm to 5 cm in diameter and may occur singly or in large numbers. The granular form appears as numerous small nodules less than 2 mm in diameter and uniformly distributed in the liver. In the diffuse form, the liver is enlarged, grayish in colour and friable. Sometimes, liver is firm, hard and gritty.

Q.2. How to differentiate MD from the lymphoid leucosis?

A.

Feature	M.D.	L.L.
Age	6 weeks and above	16 weeks and above
Symptoms	Paralysis	Nonspecific
Incidence	Usually >5% in unvaccinated herd.	Rarely >5%
Enlargement of nerves	Present	Absent
Bursa of Fabricius	Diffuse enlargement or reduction	Nodular tumours
Tumours in liver, heart, ovary, testis, skin, muscle and Proventriculus	May be present	Usually absent
Type of lymphocytes	T lymphocytes	B lymphocytes

Chapter **16**

Reticuloendotheliosis

A. Answer the followings.

Q.1. What are the different species of birds affected by REV?

A. Turkey, chicken, ducks, geese, pheasants, Japanese quails and peafowl.

B. Write true or false about the followings.

Q.1. REV can integrate into the genome of large DNA viruses including MDV and FPV.

A. True.

Q.2. Strain T of REV is defective for replication in CEF tissue culture and possess a unique oncogene of cellular origin (*v-rel*) responsible for acute oncogenicity.

A. True.

Q.3. Strain T of REV is replication defective and responsible for acute oncogenicity.

A. True.

Q.4. Non defective REV are responsible for runting diseases and chronic neoplastic disease.

A. True.

Q.5. REV is a potential contaminant of poultry vaccines.

A. True.

Q.6. The preference of the REV associated reverse transeriptase (RT) for Mn++ ion is a characteristic feature and different from other avian retrovirus.

A. True.

Q.7. Syncytia is produced in avian fibroblast culture infected with REV.

A. True.

Q.8. Chick bone marrow derived macrophages appear to be resistant to REV.

A. True.

Q.9. Both horizontal and vertical transmission is possible in REV.

A. True.

Q.10. Reticuloendo theliosis virus is under the genus– – –, subfamily– – –and family – – –.

A. Gammaretrovirus, Orthoretrovirinae, Retroviridae.

Chapter **17**

BACTERIAL DISEASES

Salmonella Infections

A. Answer the followings.

Q.1. Describe the morphology of the Salmonella bacteria.

A. They are rod shaped, Gram negative, nonsporogenic and facultative anaerobic.

Q.2. Describe the biochemical properties of *Salmonella pullorum* and *S. gallinarum*.

A. Both can ferment arabinose, dextrose, galactose, mannitol, mannose, rhamnose and xylose to produce acid with or without gas production. They do not ferment lactose, sucrose and salicin.

Q.3. How to differentiate *S. pullorum* from *S. gallinarum* on the basis of biochemical properties.

A. *S. pullorum* does not ferment dulcitol but *S. gallinarum* do. There is rapid decarboxylation of ornithine by *S. pullorum* but not by *S. gallinarum*.

Q.4. Describe the clinical signs of pullorum disease in chickens.

A. In the infected birds, there is peritonitis, arthritis and pericarditis. The lungs may be congested and liver is swollen and dark with haemorrhages. The caecum is enlarged, distended with hard, dry and necrotic material and there may be typhlitis (inflammation of caecum). Small, white, necrotic foci are found in liver, lung, heart and gizzard of the affected bird.

Q.5. What are the antibiotics of choice in Salmonella bacteria infections in birds?

A. Sulphonamide, Furazolidone, Chlortetracycline, Chloramphenicol and Neomycin.

Q.6. Describe the clinical signs of fowl typhoid disease in chickens.

A. Adult birds in acute cases show drop in egg production, ruffled feather and closed eyes. There is watery to mucoid yellow diarrhoea. The mortality may reach up to 50%. Post mortem lesions in acute cases include congestion of blood vessels under the skin. Carcasses have a septicaemia and jaundiced appearance. Liver is swollen, friable with dark red to black colour. Spleen may be enlarged and intestine contain thick, sticky and bile stained material. In chronic cases, the carcasses are emaciated and severely anaemic with focal necrosis in the heart, intestine, pancreas and liver. Pericarditis with turbid yellow fluid in the pericardial sac.

Q.7. What are the diagnostic tests to diagnose the Salmonella infections in birds?

A. Agglutination test, ELISA, FAT and PCR.

Q.8. What are the antibiotics used to treat the Salmonella infections in birds.

A. Furazolidone, enrofloxacin.

Q.9. How to diagnose the *S. pullorum* and *S. gallinarum* from other Salmonella bacteria.

A. Both *S. pullorum* and *S. gallinarum* are nonmotile but *S. enteritidis, S. typhimurium* are mortile.

Q.10. How to differentiate *S. pullorum* from *S. gallinarum* on the basis of biochemical test?

A.

Property	*S. gallinarum*	*S. pullorum*
Dextrose	Fermented with no gas	Fermented with gas
Lactose	No fermentation	Fermented with gas
Sucrose	No fermentation	Fermented with gas
Mannitol	Fermented with no gas	Fermented with gas
Maltose	Fermented with no gas	Usually not fermented
Dulcitol	Fermented with no gas	Not fermented
Ornithine	Not fermented	Fermented
Indole	Not produced	Not produced
Urea	Not hydrolysed	Not hydrolysed
Motility	Not motile	Not motile
Agglutination	+ with group D	+ with group D

B. Fill in the blanks.

Q.1. The genus Salmonella consists of more than ------ serologically distinct variants.

A. 2400.

Q.2. Salmonella can be grouped into ---- biochemically defined subgenera or -- genetically defined sp.

A. 5 ; 2.

Q.3. ----------- is the causative agent of fowl typhoid.

A. *S. gallinarum.*

Q.4. -------- is the causative agent of pullorum disease.

A. *Salmonella pullorum.*

Q.5. Both *Salmonella pullorum* and *S. gallinarum* have been placed in a single species namely ---------.

A. *S. enterica* subspecies *enterica* serovar *Pullorum* ; *S. enterica* subspecies *enterica* serovar *Gallinarum.*

Q.6. Pullorum disease is also called -----------.

A. Bacillary white diarrhoea.

Q.7. Selective enriched media for the isolation of Salmonella are ----- and -----.

A. Selenite F and tetrathionate broth. They can also grow in bismuth sulfite and McConkey agar.

Q.8. Mortality from pullorum disease occurs in ---- weeks old birds.

A. 2 to 3.

Q.9. Of 2400 serotypes of Salmonella, only --- % of these have been isolated from poultry.

A. 10.

Q.10. ----- has been the most commonly reported Salmonella serotype in humans and chicken products in UK.

A. *S. enteritidis.*

C. Write true or false about the followings.

Q.1. *Salmonella pullorum* and *S. gallinarum* are motile.

A. False. They are non-motile.

Q.2. Both *Salmonella pullorum* and *S. gallinarum* are highly host adapted and cause significant clostridial disease in chicken and turkey.

A. True.

Q.3. *S. pullorum* contains a thermostable toxins to which rodents and chicks are susceptible.

A. False. Only rodents are susceptible but not chick.

Q.4. A 85 Kbp plasmid plays a role in the virulence of *Salmonella pullorum* and *S. gallinarum* in chickens.

A. True.

Q.5. Ducks are resistant to pullorum disease and fowl typhoid.

A. True.

Q.6. Lighter breeds of chickens are more resistant than heavy breeds to pullorum disease.

A. True.

Q.7. Chicken and turkey become carrier after initial infection with pullorum disease with or without presence of lesions.

A. True.

Q.8. *S. pullorum* and *S. gallinarum* affect chicks and poults, whereas adults are susceptible to *S. gallinarum*.

A. True.

Q.9. Both horizontal and vertical transmission have been reported in pullorum disease.

A. True.

Chapter 18
Colibacillosis

A. Answer the followings.

Q.1. Describe the colony characteristics of *E. coli*.

A. *E. coli* colonies are bright pink, surrounded by a precipitate on McConkey's agar, dark green black metallic sheen on eosin methylene blue (EMB) agar and yellow on tergitol-7 agar.

Q.2. Describe the biochemical properties of *E. coli*.

A. *E. coli* produces indole, MR +, reduces nitrate to nitrite, VP and oxidase reaction negative. It can not grow in the presence of KCN, hydrolyses urea (urease negative), liquefy gelatin and grow in citrate medium. It can ferment glucose, maltose, mannitol, xylose, glycerol, rhamnose, sorbitol and arabinose with production of acid and gas. It can not ferment dextrin, starch or inositol.

Q.3. What are the different antigens present in *E. coli?*

A. Somatic (O), Capsular (K), Flagellar (H) and Pilus (F) antigens.

Q.4. How to differentiate the virulent and non-virulent *E. coli?*

A. Embryonated chicken eggs of 12 days old are inoculated with 100 cfu by allantoic route and results are noted. If there is <10% mortality, 10-29% mortality and >29% mortality within two days, it is designated as non-virulent, intermediate strain and virulent strains respectively.

Q.5. What are the diseases produced by *E. coli* in young chickens?

A. Colisepticaemia and yolk sac infection.

Q.6. What are the diseases produced by *E. coli* in adult chickens?

A. Egg peritonitis and coligranuloma. Besides, it causes synovitis, arthritis, osteomyelitis, salpingitis, panophthalmitis and localized abscesses. All these conditions are referred to as colibacillosis.

Q.7. How many pores are there on the egg shell of an egg?

A. 80,000-10,000. These are completely closed in freshly laid eggs.

Q.8. What is mushy chick disease (MCD) or omphalitis?

A. Yolk sac infection is the commonest cause of mortality in chicks during the first week of life and known as MCD or omphalitis. *E. coli* have been found in 70% cases. Other bacteria include Staphylococcus, Pseudomonas, Proteus, Clostridia, Bacillus and Enterococci.

Q.9. What is Hjarre's disease (HD)?

A. Coligranuloma or HD occurs in adult hens. It is caused by *E. coli*. The symptoms are non-specific. Affected birds are found dead. Post mortem examination shows hard, yellow nodular growth in the wall of intestine particularly caeca. Liver is hard, swollen and discoloured.

B. Write true or false about the followings.

Q.1. Colibacillosis in mammals is most often a enteric disease, whereas colibacillosis in poultry is typically a localized or systemic disease caused by virulent *E. coli*.

A. True.

Q.2. Most avian pathogenic *E. coli* (APEC) isolated from poultry are pathogenic only for birds and represent a low risk of disease for people or other animals.

A. True.

Q.3. *E. coli* 0157 :H7 is an important Shiga toxin producing enterohaemorrhagic pathogen of human.

A. True.

Q.4. Pigeons represent a natural reservoir of *E. coli* that could potentially represent a health hazard to people.

A. True.

Q.5. *E. coli* is a Gram negative, non-acid fast nonspore forming bacillus but motile with peritrichous flagella.

A. True.

Q.6. Mammalian pathogenic *E. coli* produces haemolysis on blood agar.

A, True.

Q.7. The haemolysis on blood agar is not a constant feature of Avian pathogenic *E. coli.* (APEC).

A. True.

Q.8. Inorganic acids are more effective than organic acid in inhibiting the growth of *E. coli.*

A. False. Organic acids are more effective.

Q.9. Pathogenic and non-pathogenic isolates of *E. coli* are similar in biochemical characteristics and drug sensitivities.

A. True.

Q.10. Other than ability to cause mortality in embryos or chicks, no single virulence factor has been identified that will distinguish all pathogenic from non-pathogenic isolates of *E. coli.*

A. True.

Q.11. Avian pathogenic *E. coli* tend to be more toxigenic than mammalian pathogenic *E. coli.*

A. False.

Q.12. The isolates of *E. coli* produce toxins that have effect on Vero cells called Vero toxigenic (VT).

A. True.

Q.13. Enterotoxigenic *E. coli* (ETEC) produce heat stable and/or heat labile (LT) enterotoxins which cause acute diarrhoea.

A. True.

Q.14. F1 fimbria of *E. coli* is expressed in tracheal epithelial cells.

A. True.

Q.15. P (F11) fimbria of *E. coli* is expressed in lower respiratory tract.

A. True.

Q.16. Both inactivated and live vaccines are available against *E. coli*.

A. True.

Q.17. Rough strains of *E. coli* autoagglutinate and can not be serotyped.

A. True.

Q.18. The O antigen of *E. coli* is the exotoxin.

A. False. Endotoxin.

Q.19. Normal healthy birds with intact defenses are remarkably resistant to naturally occurring *E. coli* including pathogenic strains.

A. True.

C. Fill in the blanks.

Q.1. Enterohaemorrhagic *E. coli* (EHEC) typified by 0157 produce shiga or shiga like toxins (Stx) which are also known as ––––––.

A. Shiga toxin producing *E. coli* (STEC).

Q.2. Heat labile (LT) toxin of *E. coli* is closely related to ––– ––––––-- toxin.

A. Cholera.

Q.3. *E. coli* is a common inhabitant in the intestinal tracts of poultry at concentration up to –––– /g.

A. 10^6.

Q.4. Among normal chickens, –––% of intestinal coliforms belong to potentially pathogenic serotypes.

A. 10-15.

Q.5. Dust in poultry houses may contain ––––– *E. coli*/g.

A. 10^5-10^6.

Q.6. There are currently ––––O, –––K, –––H and – ––F antigens of *E. coli*.

A. 167 ; 74 ; 53 ;17.

Q.7. K antigens of *E. coli* on the basis of heat stability can be subdivided into ————, ——————— and ———— forms.

A. L, A and B.

Chapter **19**

Fowl Cholera

A. Answer the followings.

Q.1. What are the synonyms of fowl cholera?

A. Avian cholera, avian pasteurellosis and avian haemorrhagic septicaemia.

Q.2. Describe the morphology of Pasteurella.

A. It is a Gram negative, non-motile, non-spore forming, stains bipolar and pili are present. It grows aerobically or anaerobically.

Q.3. Which media is best to grow *P. multocia*?

A. Dextrose starch agar with 5% avian serum is an excellent medium for isolating and growing *P. multocida*.

Q.4. Describe the colony characteristics of *P. multocida*.

A. The colonies produced by *P. multocida* are iridescent or blue with or without iridescence. Iridescent colonies are circular, smooth, convex, translucent, glistening and show a tendency to coalesce. As the colony ages, the colonies become larger and viscous.

Q.5. What are the different antibiotics used to treat fowl cholera in chickens?

A. Sulphonamides, Penicillin, Streptomycin, Oxytetracycline, Chlortetracycline and Erythromycin.

B. Fill in the blanks.

Q.1. ––––––– used the term fowl cholera in 1836.

A. Mailet.

Q.2. –––––––– in 1886 used the term haemorrhagic septicaemia.

A. Huppe.

Q.3. ---------- in 1990 used the term avian pasteurellosis.

A. Ligniers.

Q.4. *Pasteurella multocida,* accepted name proposed by ------ and ------- is now widely accepted.

A. Rosenburg and Merchant.

Q.5. Based on DNA homology studies, *P. multocida* has been divided into 3 subspecies namely -------, ------- and --------.

A. multocida, septica and gallicida.

Q.6. *P. multocida* subspecies *multocida* is the most common subspecies isolated from chicken and turkey.

A. True.

C. Write true or false about the followings.

Q.1. The Pasteurella bacteria will grow on meat infusion media and growth is enhanced when enriched by peptone, casein hydrolysate or avian serum.

A. True.

Q.2. Blood or serum from horse, cattle, sheep and goat inhibit the growth of *P. multocida.*

A. True.

Q.3. Blood or serum from chicken, ducks, swine and water buffalo has little or no effect on growth of *P. multocida.*

A. True.

Q.4. Pantothenic acid and nicotinamide are essential for growth of *P. multocida.*

A. True.

Q.5. Blue colonies are often isolated from birds with chronic type cholera.

A. True.

Q.6. Bacteria from smooth colonies are 3 to 4 million times more virulent than those of rough colonies in pigeons.

A. True.

Q.7. Pasteurella with iridescent type is associated with acute fowl cholera and highly virulent.

A. True.

Q.8. Pasteurella with blue type colonies are of low virulence and occur in flocks where fowl cholera is enzootic.

A. True.

Q.9. *P. multocida* does not produce gas but produces oxidase, catalase, peroxidase and a characteristic odour.

A. True.

Q.10. Unlike most of Gram negative bacteria, it is sensitive to Penicillin.

A. True.

Q.11. Fowl cholera can be transmitted through eggs.

A. False.

Q.12. Rats can act as reservoirs for *P. multocida* and may transmit fowl cholera.

A. True.

Q.13. Recovered birds can act as chronic carriers of *P. multocida* (fowl cholera).

A. True.

Q.14. Housing chickens in cages appear to reduce the chances of infection with *P. multocida*.

A. True.

Q.15. Turkeys are more susceptible than chickens to infections with *P. multocida* (and mature chickens are susceptible than young ones).

A. True.

Q.16. Endotoxins are produced by all *P. multocida*, both virulent and non-virulent.

A. True.

Q.17. Domestic geese and ducks are highly susceptible to fowl cholera.

A. True.

Q.18. Most species of farm animals may be carriers of *P. multocida.*

A. True.

Q.19. In acute case of fowl cholera in chickens, diarrhoea is initially watery and white in colour but later becomes greenish with mucous.

A. True.

Chapter 20
Ulcerative Enteritis

A. Answer the followings.

Q.1. What is the synonym of ulcerative enteritis in birds?

A. Quail disease.

Q.2. What are the different species of birds susceptible to ulcerative enteritis?

A. Chicken, pigeon, robins, turkey, pheasant, blue grouse and partridges are susceptible.

Q.3. What is the causative agent of ulcerative enteritis?

A. *Clostridium colinum.*

Q.4. What is the best medium to isolate the *Clostridium colinum*?

A. Tryptose phosphate agar with 0.2% glucose and 0.5% yeast extract. In chopped meat carbohydrate (CMC) broth good growth of *Clostridium colinum* occurs.

Q.5. Describe the biochemical characteristics of *Clostridium colinum.*

A. *Clostridium colinum* ferments glucose, mannose, raffinose, sucrose and trehalose. It weakly ferments fructose and maltose. Mannitol is fermented only by some strains. It does not ferment arabinose, cellobiose, erythritol, glycogen, inositol, lactose, sorbitol and xylose. Nitrite and indole are not produced.

Q.6. What are the clinical signs produced by *Cl. colinum* in poultry?

A. Birds may die without showing any symptoms. They are usually well muscled and fat and have feed in the crop. As the disease progresses, infected birds become listless, are hurdled with eyes partly closed and feathers are dull and ruffled. Quail often exhibit watery white faeces.

Extreme emaciation with atrophy of pectoral muscles is seen in birds affected 1 week or longer. Mortality in chickens varies from 2-10%.

Q.7. Describe the gross lesions produced by *Clostridium colinum* in quails.

A. In quails, acute haemorrhagic enteritis develops particularly in duodenum. Birds which survive several days show necrosis and ulceration in the intestine. Early lesions are characterized by small yellow foci with haemorrhagic borders. Ulcers may coagulate to form the large necrotic patch. Perforation of ulcer causes peritonitis. Liver may show large irregular yellow areas along the edges. Small yellow circumscribed foci surrounded by pale yellow halo are found in liver. The spleen may be congested, enlarged and haemorrhagic. Gross lesions are absent in other organs.

Q.8. How to treat the ulcerative enteritis in poultry?

A. Penicillin, ampicillin, tylosin, chlortetracycline, furazolidone are generally used. Bacitracin @ 0.005% and streptomycin @ 0.006% in feed have been found effective.

B. Write true or false about the followings.

Q.1. There is genetic differences in susceptibility to ulcerative enteritis in Japanese quail.

A. True.

Q.2. AGPT and FAT are generally used to diagnose ulcerative enteritis in poultry.

A. True.

Q.3. Ulcerative enteritis is most frequently seen in young birds.

A. True.

Q.4. Although, Ulcerative enteritis is found in a wide range of avian host, quails are undoubtedly among the most susceptible species.

A. True.

Q.5. Pigeons are resistant to Ulcerative enteritis.

A. False. Pigeons are susceptible to Ulcerative enteritis.

Q.6. Genetic differences in susceptibility to Ulcerative enteritis have been reported in Japanese quail.

A. True.

Q.7. Pure culture of *C. colinum* grown anaerobically are highly pathogenic for quail following oral administration.

A. True.

C. Fill in the blanks.

Q.1. Ulcerative enteritis occurs in chickens at ——————— of age, turkey at ———— age and quail at —————— of age.

A. 4-12 weeks ; 3-8 weeks ; 4-12 weeks.

Q.2. Oral administration of at least ———— viable *C. colinum* is required to produce Ulcerative enteritis in quails experimentally.

A. 10^7.

Chapter **21**

Necrotic Enteritis

A. Answer the followings.

Q.1. What is the causative agent of necrotic enteritis?

A. *Clostridium perfringens* type A or C.

Q.2. Describe the colony characteristics of *Clostridium perfringens* in blood agar.

A. *Clostridium perfringens* can be isolated on blood agar incubated anaerobically at 37ºC overnight and characterized by an inner zone of complete haemolysis and an outer zone of discolouration and incomplete haemolysis.

Q.3. Describe the biochemical properties of *Clostridium perfringens.*

A. Most of the strains ferment glucose, maltose, lactose and sucrose. They do not ferment mannitol. The principal products of fermentation are acetic acids and butyric acids. Gelatin is hydrolysed, milk digested but there is no production of indole.

Q.4. Describe the clinical signs and gross lesions of necrotic enteritis in chickens.

A. There is severe depression, decreased appetite in the affected birds. The affected birds are reluctant to move and have diarrhoea and ruffled feather. Gross lesions are confined to small intestine particularly jejunum and ileum. The mucosa is lined by a loosely or tightly adherent yellow or green layer pseudomembrane, over the time it becomes fibronecrotic enteritis with formation of diphtheritic membrane. Hepatitis characterized by tan coloured liver with necrotic foci and cholecystitis has been noticed.

Q.5. How to diagnose necrotic enteritis from ulcerative enteritis?

A. Necrotic enteritis usually confined to small intestine and there is no involvement of caecum and liver. But in ulcerative enteritis there are areas of necrosis in caeca and liver.

Q.6. What are the different antibiotics used to treat necrotic enteritis?

A. Virginiamycin, tylosin, penicillin, ampicillin, lincomycin, bacitracin and furazolidone. Tylosin, penicillin, lincomycin, bacitracin and oxytetracycline can be administered through drinking water; whereas Virginiamycin, Penicillin, Lincomycin, Bacitracin, Avoparcin and Nitrovin are added in feed.

B. Fill in the blanks.

Q.1. ———— toxin produced by type A and C and ——— toxin by type C of *Clostridium perfringens* are responsible for causing necrotic enteritis in poultry.

A. α; β .

Q.2. The coccidian sp. most commonly associated with necrotic enteritis are ————————, ————————— and ————————.

A. *Eimeria maxima, E. necatrix, E. acervulina* which inhabit in the small intestine.

C. Write true or false about the followings.

Q.1. *Clostridium perfringens* are Gram positive rods and can produce spores.

A. True.

Q.2. High level of fish meal or high levels of wheat, barley or rye in the diet can predispose to outbreak of necrotic enteritis.

A. True.

Q.3. One of the most common predisposing factor of necrotic enteritis is coccidiosis.

A. True.

Q.4. It is the intestinal coccidiosis and not caecal coccidiosis that causes necrotic enteritis.

A. True.

Q.5. *E. tenella* remains in the caecum and is not a predisposing factor for necrotic enteritis.

A. True.

Q.6. Clinical necrotic enteritis is resistant to bacitracin and lincomycin but sensitive to virginiamycin and penicillin.

A. True.

Chapter 22
Botulism

A. Answer the followings.

Q.1. What are the synonyms of botulism in poultry?

A. Limberneck and Western duck sickness.

Q.2. What are the different types of *Cl. botulinum* responsible for outbreak in poultry?

A. Almost all outbreaks in poultry are caused by type C and sometimes by A and B.

Q.3. Describe the morphology of *Cl. botulinum.*

A. Gram +, spore forming and exotoxin producing bacteria.

Q.4. Describe the signs of botulism in chickens.

A. Flaccid paralysis of legs, wings, necks and eyelids are prominent. Paralytic signs progress cranially from legs to include wings, neck and eyelids. Affected birds are reluctant to move. If forced to walk, they appear lame. Limberneck is the paralysis of neck. Death occur due to heat and respiratory failure. Up to 40% mortality has been reported.

Q.5. What is the line of treatment of botulism?

A. It can be treated with Na selenite and Vitamin A, D & E reduce mortality in broiler flocks. Antibiotic bacitracin (100g/ton in feed), streptomycin (1 g/litre in water) or periodic chlortetracycline treatment reduce the mortality from botulism.

B. Write true or false of the followings.

Q.1 Botulism in poultry is caused by endotoxin of *C. botulinum.*

A. False. It is caused by exotoxin of *C. botulinum.*

Q.2. Botulinum toxins are among the most powerful toxins known.

A. True.

Q.3. Type C spores are more easily heat inactivated than type A and B spores which are more resistant to heat than type E.

A. True.

Q.4. Type C botulism can be caused by ingestion of preformed toxin.

A. True.

Q.5. Western duck sickness is one of the most devastating diseases of waterfowl.

A. True.

Q.6. *C. botulinum* is found in the gut of normal chickens.

A. True.

C. Fill in the blanks.

Q.1. Most avian cases of botulism are caused by *C. botulinum* type – – –.

A. C.

Q.2. There are – – –cultural and – – – – – – – antigenically different toxigenic grouping of *Cl. botulinum*.

A. 4 ; 8 (A, B, Cα, Cβ, D, E, F and G).

Q.3. Human diseases are caused by type – – – and poultry diseases are caused by– – – – respectively.

A. A, B, E and F ; A, C and E.

Q.4. Chicken, turkey, pheasants and peafowl are susceptible to types –, – –, – – but not – –, – – – toxins.

A. A, B, C and E ; D , F.

Chapter 23

Tuberculosis

A. Answer the followings.

Q.1. What are the different synonyms of the tuberculosis in chickens?

A. Avian mycobacteriosis, avian tuberculosis and avian TB.

Q.2. Describe the morphology of *M. avium*.

A. It is an acid fast bacillus (rod shaped) but club like , curved and crooked forms are also seen.

Q.3. What are the different species of animals susceptible to *M. avium* infections?

A.

Cattle	Infections usually localized	Hamster	Susceptible (Intratesticular)
Cat	Highly resistant	Dog	Highly resistant
Guineapig	Relatively resistant	Mouse	Relatively resistant
Rat	Relatively resistant	Goat	Relatively resistant
Llama	Susceptible	Monkey	Susceptible
Rabbit	Readily infected	Sheep	Moderately susceptible
Swine	Readily infected		

Q.4. What are the diagnostic tests to diagnose *M. avium* in birds?

A. ELISA and rapid agglutination test.

B. Write true or false about the followings.

Q.1. *M. avium* serovars 1, 4 and 8 are isolated most commonly from AIDS patients.

A. True.

Q.2. *M. avium* serovar 1 is commonly isolated from wild birds and AIDS patients.

A. True.

Q.3. *M. avium* serovar 9 is most commonly isolated from chickens but rarely isolated from humans.

A. True.

Q.4. Most human *M. avium* infections are due to human to human infections than bird to human infections.

A. True.

Q.5. *M. avium* bacilli are motile and spore forming.

A. False. They are non-motile and non spore forming.

Q.6. In contrast to *M. tuberculosis* and *M. bovis, M. avium* grows at 25-45°C (39-45°C).

A. True.

Q.7. The colonies of *M. avium* are large if medium contains glycerin.

A. True.

Q.8. *M. avium* with smooth transparent colonies are virulent for chickens.

A. True.

Q.9. *M. avium* with smooth domed shaped or rough colonies are avirulent.

A. True.

Q.10. Generally, *M. avium* is more resistant to commonly used anti-tuberculosis drug as compared to *M. tuberculosis* and *M. bovis.*

A. True.

Q.11. Avian tuberculosis is less prevalent in young birds than older birds.

A. True.

Q.12. *M. avium* has been isolated from eggs of naturally infected chickens.

A. True.

Q.13. Hatched chicks from eggs of *M. avium* positive birds do not develop tuberculosis.

A. True.

Q.14. Boiling of eggs for 6 min and frying for 2 min is enough to kill *M. avium* bacteria of eggs.

A. True.

Q.15. Both live and kill vaccines are available against *M. avium.*

A. True.

Q.16. What is the best drugs to treat avian tuberculosis in birds?

A. A combination of isoniazid (30 mg/Kg), ethambutol (30 mg/Kg) and rifampicin (45 mg/Kg) for 18 months give good results.

C. Fill in the blanks.

Q.1. *M. avium* serovars ——————— are isolated most commonly from non AIDS patients.

A. 4, 8, 9, 16 and 19.

Q.2. Based on the pulse field gel electrophoresis, it is found that human isolates are more closely related to———— (species of animal) isolates than———(species of animal) isolates.

A. Pig ; bird.

Q.3. In US, the most common cause of avian tuberculosis in chickens is due to serovar —— and ———.

A. 1 and 2.

Q.4. *M. avium* serovar 3 has been isolated from wild birds sporadically but not from domestic birds in US.

A. True.

Q.5. *M. avium* has a total of ————— serovars.

A. 28.

Q.6. *M. avium* serovars 1to 6, 8 to 11 are called *M. avium* subspecies *avium.*

A. True.

Q.7. *M. avium* serovars 7, 12-20, 23 and 25 are *M. avium* subspecies *intracellulare.*

A. True.

Q.8. *M. avium, M. tuberculosis* and *M. bovis* can also infect psittacine birds.

A. True.

Chapter **24**

Infectious Coryza

A. Answer the followings.

Q.1. What is the causative agent of infectious coryza?

A. *Haemophilus paragallinarum.*

Q.2. How to differentiate *H. gallinarum* from *H. paragallinarum*?

A. For growth of *H. gallinarum* both factor X (hemin) and factor V (NAD : Nicotinamide adenine dinucleotide) are needed, whereas for *H. paragallinarum* only factor V is needed.

Q.3. Describe the morphology of *Haemophilus paragallinarum.*

A. *Haemophilus paragallinarum* is a Gram negative, rod or coccobacilli nonmotile bacteria. The capsule is present in virulent organisms.

Q.4. Describe the colony characters of *Haemophilus paragallinarum.*

A. The colonies produced by *Haemophilus paragallinarum* are like tiny dew drop, non-haemolytic up to 0.3 mm in diameter.

Q.5. How to differentiate *H. paragallinarum* from other avian Haemophilus organisms?

A. The failure of *H. paragallinarum* to ferment either galactose or trehalose and lack of catalase clearly separates this organism from other avian haemophilus organisms.

Q.6. Describe the biochemical properties of *Haemophilus sp.*

A. They can reduce nitrate to nitrite, ferment glucose without formation of gas, can not hydrolyse urea, gelatin negative and indole negative. There is presence of alkaline phosphatase enzyme and oxidase activity.

Q.7. Describe the clinical symptoms of infectious coryza in

chickens.

A. There is swelling of face and discharges from eyes and nostrils. Loss is due to poor growth and marked reduction in egg production (10-40%). In long run, it can turn into a chronic respiratory disease.

Q.8. What are the antibiotics used to treat infectious coryza in chickens?

A. Sulphonamide, Enrofloxacin, Norfloxacin and Esafloxacin are used. But, Erythromycin and Oxytetracyclines are two most commonly used antibiotics.

B. Write true or false about the followings.

Q.1. Infectious coryza affects primarily nasal passages of birds.

A. True.

Q.2. Chickens of all ages are susceptible but older birds are more severely affected by infectious coryza.

A. True.

Q.3. Inactivated vaccines can protect the birds against the disease (infectious coryza).

A. True. Two doses of inactivated vaccine each containing (10^8 cfu/ml) should be given one at 16 weeks of age and another 3-6 weeks after. The immunity is serotype specific.

Q.4. Live *H. paragallinarum* organisms stimulate a more protective response than inactivated vaccine but give rise to carriers.

A. True.

C. Fill in the blanks.

Q.1. There are – – – serotypes of *H. paragallinarum* namely – – – – – – – – –.

A. 3 ; A, B and C.

Q.2. In India, only serotypes – – – – – – and – – – of *H. paragallinarum* are prevalent.

A. Serotypes A and C.

Chapter 25
Gangrenous Dermatitis

A. Answer the followings.

Q.1. What are the different synonyms of the gangrenous dermatitis?

A. Necrotic dermatitis, avian malignant oedema, gas oedema disease, wing rot and in some instances as a component of blue wing disease.

Q.2. What are the etiological agents of the gangrenous dermatitis?

A. *Clostridium septicum, Clostridium perfringens type A, Staphylococcus aureus and E. coli* either singly or in combination are involved.

Q.3. Describe the biochemical properties of *Clostridium septicum.*

A. It ferments glucose, maltose, lactose and salicin. It does not ferment sucrose and mannitol. The principal products of fermentation are acetic acid and butyric acid. Gelatin is hydrolysed, milk not digested and indole not produced. Growth on egg yolk agar demonstrates an absence of lecithinase and lipase production.

Q.4. Which age groups of chickens are usually affected by gangrenous dermatitis?

A. It usually occurs in broiler at 4-6 weeks of age and in layers at 6-20 weeks of age.

Q.5. What are the clinical signs of gangrenous dermatitis in chickens?

A. Affected birds are depressed with loss of appetite, leg weakness, incordination and die within few hours. The period of illness is short (less than 24 hours). Mortality is 1-60%.

Q.6. What are the pathological lesions of gangrenous dermatitis in chickens?

A. The skin and underlying tissues under the wing between the thighs and over the ribs of the birds are dark, moist and devoid of feather due to sloughing. Deep red and swollen areas are found on the feet, legs and around the feather follicle. Blood tinged fluid under the skin is noticed. The muscles are coloured and oedematous with gas. The kidney, heart and liver are congested. Lungs are congested, oedematous and resemble a mass of dark red jelly.

Q.7. What are the antibiotics used to treat gangrenous dermatitis?

A. Penicillin, Tetracycline, Chlortetracycline, Oxytetracycline, Erythromycin and Bacitracin.

Chapter **26**

Staphylococcosis

A. Answer the followings.

Q.1. Describe the morphology of *Staphylococcus aureus.*

A. It is a Gram positive, cocci in shape and found in clusters when grow in solid media.

Q.2. What are the disease conditions produced by *Staphylococcus aureus*?

A. Arthritis, tenosynovitis, gangrenous dermatitis, yolk sac infection, subdermal abscesses (bumble foot), spondylitis and osteomyelitis, staphylococcal septicaemia, endocarditis and granuloma on liver, spleen and other organs.

Q.3. Describe the biochemical properties of *Staphylococcus sp.*

A. Staphylococci are aerobic, facultative anaerobic, β haemolytic, catalase positive, ferments glucose and mannitol and gelatinase positive.

Q.4. Describe the clinical signs of *Staphylococcus aureus* infections in chickens.

A. Early signs are ruffled feather, lameness, dropping of wings, reluctant to walk and fever. Birds with acute infection have swollen joint and are reluctant to move.

Q.5. What are the selective media used to isolate *Staphylococcus aureus*?

A. Mannitol salt or phenyl ethyl alcohol agar which inhibit Gram negative bacteria.

Q.6. What are the antibiotics used to treat *Staphylococcus aureus* infections in chickens?

A. Penicillin, Streptomycin, Tetracycline, Erythromycin, Novobiocin, Lincomycin and Spectinomycin.

Q.7. How to differentiate *Staphylococcus aureus* from *Staphylococcus epidermidis?*

A.

Characteristics	*Staphylococcus aureus*	*Staphylococcus epidermidis*
Colony pigment	+	-
Haemolysis	+	-
Coagulase	+	-
D. mannitol fermentation	+	-

B. Write true or false about the followings.

Q.1. *Staphylococcus aureus* infections most commonly occur in bones, tendon sheaths and joints specially tibiotarsal and stifle joint.

A. True.

Q.2. *Staphylococcus aureus* is the only species pathogenic for poultry.

A. True.

Q.3. In the liquid media, *Staphylococcus aureus* occurs in short chain.

A. True.

Q.4. Only coagulase positive strains of Staphylococcus are pathogenic for poultry.

A. True.

Q.5. Phages tend to be specific for *S. aureus* of poultry origin and can not be used to type strains from other species.

A. True.

Q.6. The capsular types of *Staphylococcus aureus* from chicken is mostly type 5 followed by type 8.

A. True. Type 5 (91%) and type 8 (9%).

Q.7. The capsular types of *Staphylococcus aureus* from turkey is mostly type 8 followed by type 5.

A. True. Type 8 (38%) and type 5 (33%).

Q.8. Coagulase negative strains of *Staphylococcus aureus* are thought to be non-pathogenic for poultry.

A. True.

Q.9. All avian species are susceptible to *Staphylococcus aureus* infection.

A. True.

Q.10. Most of the *Staphylococcus aureus* strains are β haemolytic and other Staphylococci are usually nonhaemolytic.

A. True.

Q.11. *Staphylococcus aureus* produces enterotoxin capable of food poisoning in humans.

A. True.

Q.12. Protein A, a component of cell wall of *Staphylococcus aureus* intereacts non-specifically with Fc portion of immunoglobulin and may be a virulence factor.

A. True.

Q.13. Neither active nor passive immunity is important in preventing *S. aureus* infections in poultry.

A. True.

C. Fill in the blanks.

Q.1. About – – – typical and – – – atypical *S. aureus* produce food poisoning in human beings.

A. 50 ; 50.

Q.2. The genus Staphylococcus contains approximately – – – – – – species.

A. 20.

Q.3. Staphylococcus most frequently isolated from poultry include – – –, – – and – – –.

A. *S. aureus* and *S. epidermidis* from poultry and *S. gallinarum* from processed poultry.

Chapter **27**

Streptococcosis

A. Answer the followings.

Q.1. Describe the morphology of Streptococci.

A. They are Gram +, spherical and occur in pair or short chain, non-motile, non-spore forming and facultative anaerobe. They are catalase negative and ferment sugars usually to lactic acid.

Q.2. Describe the clinical signs of Streptococci infections in chickens.

A. It occurs in two form. In acute form, there is marked depression, ruffled feather, pale comb and wattles, decreased egg production, head tremors and diarrhoea. In chronic form, there is loss of body weight, lameness, diarrhoea, fever, head tremor and torticollis.

Q.3. Describe the gross lesions of Streptococci infections in chickens.

A. In chronic form, there is dehydration, emaciation, arthritis, peritonitis, endocarditis, pericarditis and pneumonia. The heart is enlarged and there may be ascites.

Q.4. How to diagnose the Streptococcus infections in chickens?

A. Isolation of streptococcus in blood agar and latex agglutination test.

Q.5. What are the antibiotics used to treat the Streptococci infections in chickens?

A. Penicillin, Erythromycin, Novobicin, Oxytetracycline, Chlortetracycline, Tetracycline and Enrofloxacin.

B. Write true or false about the followings.

Q.1. *S. zooepidemicus* is also called *S. gallinarum*.

A. True.

Q.2. *S. faecium* and *S. mutans* cause mortality in duckling and goslings respectively.

A. True.

C. Fill in the blanks.

Q.1. Among the Streptococci, ---- and -- species are commonly occurs in chickens.

A. *S. zooepidemicus* in mature chickens and *S. faecalis* in chickens of all ages.

Q.2. *S. zooepidemicus* is under Lancefield serogroup type ---.

A. C.

Q.3. *S. faecalis* is under Lancefield serogroup type ---.

A. B.

Chapter 28

Campylobacteriosis

A. Answer the followings.

Q.1. What is the synonym of campylobacteriosis in chicken?

A. Avian vibronic hepatitis.

Q.2. What are the different species of Campylobacter pathogenic for poultry?

A. *Campylobacter jejuni, C. coli* and *C. laridis.*

Q.3. Describe the morphology of *Campylobacter sp.*

A. It is a Gram positive, microaerophilic, spiral uniflagillate organism which show characteristic darting motility when examined under dark field microscope.

Q.4. What are the different media used to isolate and grow *Campylobacter sp.*?

A. Medium- BU 40, Skirrow's medium, Butzler's medium, Blaser's medium and CAMPY -BAP medium.

Q.5. Describe the morphology/colony characteristics of *Campylobacter sp.*

A. The colonies are either flat, translucent with a tendency to coalesce or raised, opaque and brown gray with discrete margins. Colonies are non haemolytic on blood agar.

Q.6. Describe the biochemical properties of *Campylobacter sp.*

A. As Campylobacters are unable to ferment carbohydrate, energy is derived from amino acids. Three thermophilic species of clinical significance reduce selenite, oxidase and catalase positive and indole negative.

Q.7. How to differentiate different species *Campylobacter sp.*

A.

Species	Grow at 25°C	Grow at 42°C	Nalidixic acid sensitivity	Hippurate hydrolysis
C. foetus	+	-	R	-
C. *coli*	+	+	S	-
C. jejuni	-	+	S	+
C. *laridis*	-	+	R	-

Q.8. How to differentiate different biotypes of *Campylobacter sp.*?

A.

Properties	*C. jejuni* biotypes				*C. coli* biotypes		*C. laridis* biotypes	
	I	II	III	IV	I	II	I	II
Hippurate hydrolysis	+	+	+	+	-	-	-	-
Rapid H_2S test	-	-	+	+	-	-	+	+
DNA hydrolysis	-	+	-	+	-	+	-	+

Q.9. What are the antibiotics used to treat Campylobacteriosis in chickens?

A. Erythromycin, doxycycline, chloramphenicol, furazolidone and gentamycin.

Q.10. What are the clinical signs and pathological lesions of Campylobacteriosis in chickens?

A. In chickens, symptoms are limited to depression and diarrhoea. In broiler, there is reduced weight gain and poor food efficiency. In layers, there is decreased egg production. The gross lesions include distension of intestinal tract from duodenum to the bifurcation of the caeca. Accumulation of mucous and watery fluid may be present. The newly hatched chicks show red or yellow mottling of the liver.

Q.11. What are the antibiotics used to treat Campylobacteriosis in chickens?

A. Erythromycin, doxycycline, chloramphenicol, furazolidone and gentamycin.

B. Write true or false about the followings.

Q.1. All the *Campylobacter sp.* are motile with a single polar flagellum.

A. True.

Q.2. Cockroaches and house flies can be carrier of *C. jejuni* and play a role in transmission of the disease.

A. True.

Chapter 29

Nutritional Diseases

A. Answer the followings.

Q.1. What are the different amino acids absolutely essential for poultry?

A. 10 absolutely essential amino acids are arginine, histidine, isoleucine, leucine, lysine, methionine, phenylalanine, threonine, tryptophan and valine. Two amino acids namely cysteine and tyrosine can be synthesized from essential amino acids. Two amino acids that are essential for young chicks are glycine or serine and proline.

Q.2. What are the essential fatty acids for poultry?

A. Linoleic acid and arachidonic acid.

Q.3. What is encephalomalacia or crazy chick disease?

A. Encephalomalacia or crazy chick disease is caused by deficiency of Vit E. It is a nervous disorder which is characterized by muscular weakness, muscular incordination, torticollis, backward and downward retraction of head, rapid contraction and relaxation of the legs, paralysis and death.

Q.4. What are the different conditions occur in poultry due to Vit E deficiency?

A. (i) Exudative diathesis - It is an oedema of the S/C tissue under the skin. (ii) Nutritional muscular dystrophy (myopathy) - it occurs when Vit -E deficiency is accompanied by a deficiency of S containing amino acids methionine and cysteine.

Q.5. What are the different nutrient deficiencies cause greater incidences of disease in birds?

A. Linoleic acid, Vitamin A, E, B complex, Iron, Se.

Q.6. Arrange the amino acids in the decreasing order of toxicity in poultry.

Methionine, Tryptophan, histidine, phenylalanine, threonine, tyrosine, lysine, leucine, valine, arginine and isoleucine.

A. Methionine > Phenylalanine > Tryptophan = Histidine = Lysine > Tyrosine = Threonine > Isoleucine > Arginine > Valine = Leucine.

B. Fill in the blanks.

Q.1. Lack of fatty acid in the poultry ration causes ------------- symptoms.

A. Enlarged fatty liver.

Q.2. All the vitamins except vitamin --- is dietary requirement for poultry.

A. C.

Q.3. Chondrodystrophy, enlargement of hock joints and short, thick and usually deformed long bone is caused by deficiency of -------.

A. Mn and deficiency of any one of the following vitamin choline, nicotinic acid, pyridoxine, biotin and folic acid.

Q.4. Vit D deficiency causes ------ in young chickens and ------ in adult birds.

A. Rickets ; Osteoporosis.

Q.5. Vit K was discovered in chickens by ------ in the year -------- and received Nobel prize.

A. Henrik Dam ; 1943.

Q.6. ----------- in 1929 discovered Vit B1 (thiamine) in chickens and received Nobel prize.

A. Christian Eijkman.

Q.7. Vit B1 deficiency causes ---------- in poultry.

A. Polyneuritis, star grazing (opisthotonus).

Q.8. The dose of Vit B1 in young chickens is ---mg/day and in older birds--mg /day.

A. 5-10 ; 10-50.

Q.9. The requirement of Vitamin B_2 is – – mg/chick and – – – mg/layer per day.

A. 5 ;15.

Q.10. Pantothenic acid is also called vitamin – –.

A. B_5.

Q.11. The dose of the pantothenic acid is – – – – mg/chick/ day.

A. 10-20.

Q.12. Choline is also called Vitamin – – or – – –.

A. B4 ;B7.

Q.13. – – – – – – – (sp.of bird) synthesizes Vitamin C only in kidney, – – – – – – (sp.of bird) only in liver and – – – – both in liver and kidney.

A. Chicken ; sparrow ; crow.

Q.14. Macromolecules essential for chicken are –.

A. Ca, P, Mg, K, Na and Chlorine.

Q.15. Trace elements (Microminerals) essential for chicken are –.

A. Mn, Fe, Cu, Zn, I_2, Mb and Se.

Q.16. The ratio of Ca and Phosphorus in bone is – – – – – – – – – – – in chickens.

A. 2.5 : 1.

Q.17. The ratio of Ca and Phosphorus in chicken's egg is – – – – – – – – – – –.

A. 20 :1.

Q.18. Ca and P deficiency causes – – – – – – – – – in young growing bird and – – – – – – – – in layer and growing bird.

A. Ricket; Osteoporosis (cage layer fatique).

Q.19. The lethal dose of NaCl in chickens is – – – – – – /Kg Bwt.

A. 4g.

Q.20. Excess Na in the ration results in ––––––– and ––––– in broiler chicken.

A. Ascites ; visceral gout.

Q.21. Egg shells contain about –––––––% Mg.

A. 0.4.

Q.22. Mn deficiency causes ––––––––––––––– in growing birds.

A. Chondrodystrophy.

Q.23. Thyroxine, an –––––––(metal) containing hormone synthesized by thyroid.

A. Fe.

Q.24. Most of the blood Cu is found as a component of –––––––, a Cu binding protein with enzymatic activity.

A. Ceruloplasmin.

Q.25. Haemoglobin is an ––––––(metal) containing pigment of RBC.

A. Fe.

Q.26. Selenium is a constituent of cellular enzyme –––––––––––.

A. Glutathione peroxidase.

Q.27. No toxicity will be observed if zinc is added @ –––– g/Kg of poultry feed.

A. 1.

Q.28. The selenium should be added for chickens at the rate of –––––– ppm upto 16 weeks of age in the ration.

A. 0.1 ppm.

Q.29. Zinc is a co-factor of more than –––– enzymes in the body.

A. 200.

Q.30. Grains such as –––––––, –––––––– and –––––– are more deficient in lysine.

A. Corn ; Milo ; Wheat.

Q.31. ––––––––––meal is most deficient in methionine.

A. Soya bean.

Q.32. ----------amino acid deficiency causes impaired pigmentation of bronze turkey poults and stunting and retarded development in chicks.

A. Lysine.

Q.33. ---------- amino acid deficiency causes wing feather to curl upward.

A. Arginine.

C. Write true or false about the followings.

Q.1. When Soya bean is used in the poultry ration as protein source, methionine supplementation is necessary.

A. True.

Q.2. Vit D3 (Cholecalciferol) is essential for poultry, human and animals.

A. True.

Q.3. Vit D2 (Cholecalciferol) is essential for human and animals but not for poultry.

A. True.

Q.4. Vit D is essential for proper metabolism of Ca and P in the formation of bones.

A. True.

Q.5. Vitamin B_2 (Riboflavin) is not stored in the body but stored in the yolk of the eggs.

A. True.

Q.6. Vitamin B_{12} is synthesized in the adult birds not in young chicks by bacteria in the caeca and large intestine.

A. True.

Q.7. Vitamin B_2 deficiency cause curled toe paralysis in chickens.

A. True.

Q.8. Clubbed down, a condition where in proper growth of down (soft feathers with a club shaped appearance) occurs due to Vitamin B_2 deficiency.

A. True.

Q.9. Niacin is synthesized from tryptophan in the body of the chickens.

A. True.

Q.10. Pantothenic acid is an essential part of Co-enzyme A.

A. True.

Q.11. Pantothenic acid is also called chick anti-dermis factor.

A. True.

Q.12. Biotin deficiency causes fatty liver and kidney syndrome in chickens.

A. True.

Q.13. During choline deficiency in birds, triglycerides do not converted into very low density lipoprotein and accumulate in the liver called fatty liver.

A. True.

Q.14. Most of the animals and birds can synthesize sufficient amount of Vitamin C from glucose.

A. True.

Q.15. Chickens can synthesize Vitamin C both in kidney and liver.

A. False. Only in kidney not in liver.

Q.16. Most of the birds synthesize Vitamin C only in the liver.

A. True.

Q.17. Zinc is an essential trace element for poultry.

A. True.

Q.18. Excess zinc is excreted through pancreatic secretion and bile in birds.

A. True.

Q.19. Vitamin A deficiency reduces antibody responses and depletion of lymphocytes in thymus and bursa.

A. True.

Q.20. Sodium selenite can supply Se @ 0.1 ppm when 464 g added to 1023 Kg of feed.

A. True.

Q.21. White Leghorn with Vitamin E or Se deficiencies have decreased humoral immune response to low immunogen dose but not with high immunogen dose.

A. True.

Q.22. Chicks deficient in Vit E and Se have impaired development of the bursa, spleen and thymus.

A. True.

Q.23. Both excess dietary protein and deficiencies of amino acids result in increased water intake in birds.

A. True.

Q.24. Methionine is the most toxic among all the amino acids for poultry.

A. True.

Q.25. Fats are source of essential nutrients linoleic acid and arachidonic acid.

A. True.

Q.26. Methionine deficiency may exacerbate choline or Vitamin B_{12} deficiency.

A. True.

Chapter 30

Metabolic Diseases

A. Answer the followings.

Q.1. What is gout?

A. Gout is a condition in which high levels of uric acid is present in the blood (hyperuricaemia). Gout occurs in two forms (i) visceral (ii) articular. In the visceral form, monosodium urates deposit in the kidney, mesentry, peritoneum, heart, liver and spleen. The deposits on the heart, liver, airsac and peritoneum appear as white chalky coating. The articular gout is characterized by Tophi. Tophi are deposits of urates around joints, enlargement of joints and feet appear deformed.

Q.2. Why are birds more prone to develop gout compared to mammals?

A. Birds are uricotelic. The nitrogenous by-product of protein metabolism is in the form of uric acid. Birds also lack the enzyme uricase which converts uric acid into allantoin. Again, uric acid is water insoluble. Any injury or damage to kidney interferes with the elimination of uric acid, accumulates in the blood and causes visceral gout. On the other hand, mammals are ureotelic. Nitrogenous by-product of protein metabolism is urea which is water soluble. That is why mammals are not prone to gouts like birds.

Q.3. What are the different symptoms observed in chickens at different level of pyridoxine in the poultry ration?

A.

Level of Pyridoxine in the diet	Level of protein in the diet	Symptoms/Diseases
Low level of pyridoxine (2.2 mg /Kg diet)	High level of protein (31%)	Classic nervous signs
Intermediate level of pyridoxine (2.5-2.8 mg /Kg diet)	High level of protein (31%)	Chondrodystrophy but no nervous signs (bone curvature)
Lowest level of pyridoxine (1.9 mg /Kg diet)	22% protein	No nervous signs, no Chondrodystrophy, no lower growth rate

Q.4. How does alkalis reduce ascites in broiler?

A. The addition of 1% $NaHCO_3$ in feed causes alkalosis which in turn reduces ascites. This is because pulmonary constriction is mediated by H+ concentration and OH- neutralize this in alleviating the ascites.

Q.5. How does L-arginine supplementation in feed reduce mortality from ascites in birds?

A. L-arginine is required as a substrate for the production of nitric oxide, a powerful vasodilator. Nitric oxide relaxes the tone of the vessels and permits increased blood flow.

B. Write true or false about the followings.

Q.1. Vitamin A deficiency can also cause outbreaks of visceral gout in poultry.

A. True.

Q.2. The left ventricle of the heart in the bird is very thin walled, whereas the right ventricle is thick walled.

A. False. Left ventricle – thick walled ; right ventricle – thin walled.

Q.3. Linoleic acid can not be synthesized in the body of the chickens but can be converted to arachidonic acid.

A. True.

Q.4. Niacin requirement in chicks and hens depend on the level of tryptophan in the diet.

A. True.

Q.5. Light breeds of the Mediterranean class have been much more prone to feather pecking than heavier breeds of the American and Asiatic classes.

A. True.

Q.6. Vitamin C, Vitamin E and organic Se are effective in reducing mortality from ascites in birds.

A. True.

Q.7. The RBC of broiler are more rigid than those of layers.

A. True.

Q.8. Eggs from older hens have more biotin than younger ones.

A. True.

C. Fill in the blanks.

Q.1. ————————————, a mycotoxin cause damage to kidney and extensive visceral gout.

A. Oosporein.

Q.2. Bird's heart is —————— shaped with ——————— chambers.

A. Cone ; 4.

Q.3. All recognized Vitamins except Vit———— is dietary requirement of poultry.

A. C.

Q.4. Chondrodystrophy (Perosis) occurs in young chicks or poults due to deficiency of ———— (metal) and any one of the —————, ———————, ——————, —————— or ————— Vitamins.

A. Mn ; Choline, Nicotinic acid, Pyridoxine, biotin or folic acid.

Q.5. Se at the level of ————————ppm prevent or cure exudative diathesis in Vit E deficiency.

A. 0.04 -0.1.

Q.6. Se at the level of -------- ppm effectively prevents myopathies of ventricles and heart in young poults.

A. 0.1-0.2.

Q.7. Niacin (Nicotinic acid) is the Vitamin component of two important co-enzymes namely----------- and ---------.

A. NAD and NADP (Nicotinamide adenine dinucleotide phosphate).

Q.8. When tryptophan is marginally adequate in the ration, chicken can synthesize about 1 mg of Niacin from ----- mg of dietary tryptophan.

A. 45.

Q.9. Duck can synthesize 1 mg of Niacin from ----mg of dietary tryptophan.

A. 180.

Q.10. Pyridoxine is required for function of ----- and ------ enzymes.

A. Pyridoxal phosphate ; pyridoxamine phosphate.

Q.11. Average liver fat in different strains of layer may vary from ---- to ---%.

A. 25 to 50.

Chapter 31

Mycotoxins

A. Answer the followings.

Q.1. What are the different organic acids used as mycotoxin inhibitors?

A. Proprionic acid, acetic acid, citric acid, sorbic acid and benzoic acid.

Q.2. What are the different substances other than organic acids can be used as probiotics?

A. Tetrasodium pyrophosphate, alkaline polyphosphate, ammonium hydroxide, essential oil extracts, Potassium sorbate, silicon dioxide, propylene glycol, 3-p-cymenol and oxine copper.

Q.3. What are the different toxins bound by the MOS (Mannan oligosaccharide)?

A. It can bind to aflatoxin, ochratoxin, T-2, fumonisin, citrinin etc.

B. Write true or false about the followings.

Q.1. Mould inhibitors have no action on already formed toxin.

A. True.

Q.2. Mould inhibitors are added to feed only to prevent the fungal growth.

A. True.

Q.3. Mycotoxin inhibitors can destroy Vit E in the feed.

A. True.

Q.4. Hydrated sodium calcium alumino silicate (HSCAS) binds only aflatoxin.

A. True.

C. Fill in the blanks.

Q.1. Ergotism in poultry caused by ergot is produced by – – – – – – – – – – – – – – – – –.

A. *Claviceps purpurea.*

Q.2. – – – – – – – acid is the chemical building block of 40 or more alkaloids produced by *Claviceps species.*

A. Lysergic acid.

Q.3. Zeolytes, the toxin binders reduce the effect of – – – – mycotoxin but not – – – mycotoxin.

A. Aflatoxin ; Ochratoxin.

Q.4. Activated charcoal binds with – – – – and – – – – – – – mycotoxins.

A. T2 and Aflatoxin.

Chapter **32**

Vaccines and Vaccination

A. Answer the followings.

Q.1. What should be the volume of the water in the drinking water vaccination?

A.

Age of birds (in weeks)	Water (Litre/1000 doses)
4-5	15
15-16	20-25
Adults	30-50

Q.2. What is the schedule of the broiler vaccination?

A.

Day	Vaccine	Route
1	M.D.	S/C or I/M
6-7	RD (F1 or B1)	Eye drop/I/N
10-12	Gumboro (intermediate)	Drinking water
18-21	Lasota	Drinking water
24-30	Gumboro (intermediate)	Drinking water

Q.3. What is the schedule of the layer vaccination?

A.

Day/week	Vaccine	Route
1	MD	S/C or I/M
1-3	Gumboro (intermediate)	Eye drop
7	Lasota	Eye drop
14	Gumboro (intermediate) (Repeat)	Eye drop

Day/week	Vaccine	Route
18	MD	0.2 ml I/M or S/C
21-23	IB + Lasota	Drinking water
28-30	IBD (Intermediate) (Repeat)	Drinking water
42	Fowl pox	Wing web
Week 8	Lasota (Repeat)	Drinking water
	Infectious Coryza	I/M
Week 11-12	IB + Lasota (Repeat)	Drinking water
13	RD (R2B)	I/M
14	Fowl pox (Repeat)	Wing web
18	RD (killed)	S/C

B. Write true or false about the followings.

Q.1. If Avian encephalomyelitis and IB H52 strain vaccine given to old and unvaccinated bird, they may produce disease.

A. True.

C. Fill in the blanks.

Q.1. In spray vaccine, aerosol particles of – – – – µM is preferred.

A. 10.

Chapter 33
Miscellaneous

A. Answer the followings.

Q.1. What is probiotic?

A. Probiotic is the substances that promote life, whereas antibiotic is a substance that acts against life. At present, a probiotic means a live microbial feed supplement.

Q.2. How do the lactic acid bacteria act as probiotic?

A. Lactic acid bacteria produce several bactericidal substances namely bacteriocins, lactocidin, lactic acid, acetic acid, H_2O_2, lysozyme, lactoferrin, lactoperoxidase and acidophilin and check the growth of bacteria.

Q.3. How does *Saccaharomyces cerevisiae* act as probiotic?

A. *Saccaharomyces cerevisiae* releases mannan oligosaccharide (MOS) which is antagonistic to disease producing organisms.

Q.4. What are the different bacteria used as probiotic?

A. *Lactobacillus acidophilus, L. casei, L. bulgaricus, Streptococcus faecium and Bacillus subtilis.*

B. Write true or false about the followings.

Q.1. *Lactobacillus species* (bacteria) can be used as probiotic.

A. True.

Q.2. Enterotoxin released by *E. coli* can be neutralized by a metabolite produced by *L. bulgaricus.*

A. True.

C. Fill in the blanks.

Q.1. *Lactobacillus sp.* secrete ———————— enzyme which help in carbohydrate digestion.

A. α amylase.

Q.2. *Lactobacillus sp.* secrete -------- enzyme which help in protein digestion.

A. Protease.

Q.3. *Lactobacillus sp.* secrete -- ----- enzyme which help in fat digestion.

A. Lipase.

Q.4. The carcasses of birds died of heat stress lose the normal colour and become pale called --------.

A. Cooked meat appearance.

Q.5. Ammonia level in the poultry house should be maintained less than ------- ppm to avoid any harmful effect.

A. 25.

Q.6. Bumble foot in chicken is caused by -----------.

A. Staphylococci bacteria.

Chapter **34**

PARASITIC DISEASES

Ectoparasites

A. Write true or false about the followings.

1. Mites are very small parasites, barely visible to the naked eye, which live by sucking the blood of the bird.
2. Mites can not live for several months without food, so destocking a shed for a short time can eliminate the problem.
3. *Argas persicus* is commonly known as red mite or roost mite of poultry.
4. Inspection of birds during day time may not reveal infestation of the red mites.
5. The whole life cycle of *Dermanyssus gallinae* can be completed in 7 hours.
6. *Ornithonyssus (Liponyssus)* mites leave the host to lay eggs in cracks and crevices and when they hatch the larvae move about in search of a bird.
7. Bed bugs (*Cimex lectularius*) are annoying insects in human dwellings, at times it causes severe irritation and anaemia in poultry.
8. Life cycle of lice undergo ecdyses in which eggs develop to larvae, the nymph and then adult by metamorphosis.
9. Generally, lice lives only on the host while mites lives on host and in environment.
10. *Menopon gallinae is* commonly known as wing louse of poultry.
11. The pigeon fly (hippoboscid fly, louse fly), *Pseudolynchia canariensis,* is especially harmful to nestling pigeons and also transmits *Haemoproteus columbae* which causes a

malarial-like disease of pigeons.

12. The pigeon fly is unusual in that the larvae matures inside the female and pupates immediately upon ejection.

Ans.

1. True
2. False. They can live for several months, so destocking for a short time can not eliminate.
3. False. *Argas persicus* is commonly known as common fowl tick.
4. True
5. False. In 7 days.
6. False. Eggs are laid largerly on the host.
7. True
8. False. No larval stage is seen and there are little or no metamorphosis.
9. True
10. False. It is known as shaft louse.
11. True
12. True

B. Match the following.

1.	Northern mite of poultry	a) *Knemidocoptes gallinae*
2.	Tropical fowl mite	b) *Argas persicus*
3.	Subcutaneous Mite	c) *Echidnophaga gallinacea*
4.	Bed bugs	d) *Ornithonyssus bursa*
5.	Red mite of poultry	e) *Cytodites nudus*
6.	Common fowl tick	f) *Dermanyssus gallinae*
7.	Air sac mite	g) *Knemidocpotes mutans*
8.	Depluming itch mite	h) *Laminosioptes cysticola*
9.	Sticktight flea	I) *Cimex lectularius*
10.	Scaly leg mite	j) *Ornithonyssus sylviarum*

Ans. 1. j 2. d 3. h 4. I 5. f 6. b 7. e 8. a 9. c 10. g

C. Fill in the blanks.

1. Red Mite (*Dermanyssus gallinae*) of poultry feed only at .... time.
2. During the .... time, red mites hide in dark cracks, crevices or nesting materials.
3. .... is probably the commonest louse found in chickens.
4. *Cimex lectularius*, commonly known as .... , can survive without food for over a ....
5. Species of the order Phthiraptera are commonly known as ....
6. All the poultry lice belong to sub order .... , commonly called .... lice.
7. In lice, the tarsi of species parasitic on mammals have .... claw, whereas those of parasitic on birds have .... claw.
8. The sticktight flea, ... , is found on a wide variety of birds and mammals. The adults are usually attached to the skin of the .... of the host in clusters of up to ....
9. The sticktight flea is unique among poultry fleas as the .... are deeply embedded in the skin making the adults sessile while other fleas of birds and mammals are .... feeders on poultry.
10. The .... fly (hippoboscid fly, louse fly), *Pseudolynchia canariensis*, is especially harmful to nestling pigeons and also transmits .... which causes a malarial-like disease of pigeons.

Ans. 1.night 2. day 3. *Menacanthus stramineus* 4. bed bug, month 5. lice 6. Mallophaga, biting 7. one, two 8. *Echidnophaga gallinacea*, head, 100 9. mouthparts, intermittent 10. pigeon, *Haemoproteus columbae*

D. Match the following lice of poultry with their scientific names:

1.	Shaft louse	A. *Lipeurus caponis*
2.	Body louse	B. *Menopon gallinae*
3.	Wing louse	C. *Goniocotes gallinae*
4.	Head louse	D. *Menacanthus stramineus*
5.	Fluff louse	E. *Chelopistes meleagridis*
6.	Large turkey louse	F. *Cuclogaster heterographus*

Ans. 1. B 2. D 3. A 4. F 5. C 6. E

Chapter 35

Cestoda and Trematoda

A. Match the following.

1. Proglottid — A. Sporocyst, redia, cercaria
2. Oral sucker — B. *Choanotaenia infundibulum*
3. Microscopic Tapeworm — C. Anti cestodal compound
4. Large tapeworm — D. Metacercaria
5. Slender thread like tapeworm — F. Head of tapeworm
6. Life cycle of trematode — G. Cysticercoid
7. Life cycle of cestodes — H. *Hymenolepis carioca*
8. Infective stage of poultry cestodes — I. Trematode
9. Infective stage of trematodes — J. *Davainea proglottina*
10. Scolex — K. Individual segments of tapeworm
11. Praziqantel — L. Onchosphere

Ans. 1. K 2. I 3. J 4. B 5. H 6. A 7. L & G 8. G 9. D 10. F 11. C

B. Fill in the blanks.

1. Cestodes and trematodes are come under phylum .... , dorsoventrally .... without body cavity.
2. The life history of both cestodes and trematodes is usually .... , requiring one or two .... host.
3. Tapeworms are hermaphrodite, segmented body without an .... canal, belong to class ....,

whereas, flukes body are not segmented and consist of one piece only, belong to class ....

4. Among the Raillietina species, ... is the most pathogenic. In this infection, ... are formed at the site of attachment and a .... enteritis may occur.

5. *Davainea proglottina* occurs in the .... loop of the small intestine. This tapeworm is measured about .... mm long and contain 4-9 segments.

6. Birds become infected with *Davainea proglottina* by ingestion of infective .... (intermediate host) containing .... (infective stage).

7. More than .... species of tapeworms have been described from domestic and wild birds.

8. *Davainea proglottina* generally shed .... Gravid proglottid per day, whereas, *Raillietina cesticillus* may shed as many as .... per day.

9. Metacestode is the term used for .... tapeworm which develop in the .... host.

Ans. 1. Platyhelminthes, flattened 2. indirect, intermediate 3. alimentary, Cestoda, Trematoda 4. *Raillietina echinobothrida*, nodules, hyper plastic 5. duodenal, 0.5 to 3, 6. Slugs, cysticercoid 7. 1400 8. One, 10 to 12
9. larval, intermediate

C. Match the following

1 *Prosthogonimus pellucidus*	a) Pseudoscolex
2 *Davainea proglottina*	b) Hyperplastic enteritis
3 *Metroliasthes lucida*	c) Ostrich
4 *Fimbriaria fasciolaris*	d) Stable fly or beetles
5 *Raillietina echinobothrida*	e) Bursa of Fabricius
6 *Collyriclum faba*	f) Grasshopper
7 *Houttuynia struthionis*	g) Occurs in caeca and rectum
8 *Amoebotaenia cuneata*	h) smallest tapeworm of poultry
9 *Echinostoma revolutum*	i) Subcutaneous tissues

10 *Hymenolepis carioca* j) Earth-worm

Ans. 1. e 2. h 3. f 4. a 5. b 6. i 7. c 8. j 9. g
10. d

D. Write true or false about the following statements.

1. The life cycles of cestoda and trematoda are usually direct.
2. As *Davainea proglottina* is the smallest tapeworm, so it is less pathogenic.
3. Earth-worm acts as intermediate host of *Davainea proglottina.*
4. Trematodes are flattened, ribbon shaped and segmented worms.
5. Cestodes (tapeworms) are commonly encountered in poultry from free-range or backyard flocks.
6. Cysticercoids are developed only invertebrate hosts.
7. Most of the cestodes have no common name; thereby they are best recognized by their genus and species name.
8. *Raillietina echinobothrida* is usually listed as one of the most pathogenic tapeworm of poultry.
9. For cestodes and trematodes, control measures are best directed against the intermediate host.
10. Most of the poultry tapeworms belong to family Taeniidae.
11. All tapeworms of birds occurs in small intestine.
12. Infective stage of all poultry tapeworms is cysticercoid.

Ans. 1. False. Indirect; 2. False. It is most pathogenic; 3. False. Molluses for *Davainea proglottina,* earthworm for *Amoebotaenia sp.* 4.False. They are estodes (tapeworms) 5. True; 6. True; 7. True; 8. True; 9. True; 10. False; 11. True; 12. True

E. Identify and name the tapeworm species of birds of the following features:

1. This microscopic tapeworm has only 4-9 proglottids and is 0.5-3 mm. long.

A. *Davainea proglottina*

2. It is one of the largest (up to 25 cm long) and common

fowl tapeworm with cosmopolitan distribution and it belongs to genus *Raillietina.*

A. *Raillietina tetragona*

3. The most commonly found member of genus *Raillietina* and can be recognized easily from other members by the absence of neck and a large scolex bearing wide rostellum.

A. *Raillietina cesticillus*

4. The young worms penetrate with their anterior and deeply into the mucosa and submucosa of the duodenum resulting formation of nodules at the site of their attachment.

A. *Raillietina echinobothrida*

5. One of the commonest species found in the fowl under family Hymenolepididae.

A. *Hymenolepis carioca*

6. It occurs in ducks and geese and is one of the most harmful parasite of Hymenolepididae family.

A. *Hymenolepis lanceolata*

7. The anterior portion of the body forms a folded expansion or pseudoscolex.

A. *Fimbriaria fasciolaris*

8. Externally the body is finely segmented, but this does not correspond to the internal arrangements of the organs.

A. *Fimbriaria fasciolaris*

9. Birds are infected with the parasites by eating of copepods.

A. *Hymenolepis lanceolata and Fimbriaria fasciolaris*

10. Small worm with elongate triangular shape, rarely over 4 mm in length, containing about 20 proglottids.

A. *Amoebotaenia sphenoides*

11. Grasshoppers act as intermediate host of this rare tapeworm of which eggs pass into a large paruterine organs.

A. *Metroliasthes lucida*

12. Large tape worm of fowl, up to 107mm long, having two sets of genital organs in each segment.

A. *Cotugnia digonopora*

13. It occurs in duck in Burma of genus *Cotugnia*.

A. *Cotugnia fastigata*

14. It occurs in pigeons in Burma and India of genus *Cotugnia*.

A. *Cotugnia cuneata*

15. This parasite grow up to 60 cm long and 9 mm wide which occurs in the small intestine of ostrich and the affected chicks often die in severe infection.

A. *Houttuynia struthionis*

16. This large robust tapeworm of fowl and turkey is extremely white and is readily seen attached to the upper half of small intestine. The eggs of it, are with distinctive elongated filaments.

A. Choanotaenia infundibulum.

F. Match the following.

1. Oviduct fluke	a) *Tracheophilus cymbius*
2. Double pore tapeworm of poultry	b) *Echinostoma revolutum*
3. Head collar around oral sucker	c) *Davainea proglottina*
4. Bird flukes of air sac/nasal cavity	d) *Prosthogonimus* spp.
5. Most pathogenic tapeworm of poultry	e) *Raillietina echinobothrida*
6. Nodule formation at site of attachment	f) *Cotugnia digonopora*

Ans. 1. d 2. f 3. b 4. a 5. c 6. e

G. Answer the followings in brief.

1. Write two important tapeworm parasites of turkeys.

A. *Raillietina georgiensis* and *Metroliathes lucida*.

2. Write two important tapeworm parasites of ducks and geese.

A. *Fimbriaria fasciolaris* and *Hymenolepis megalops*.

Chapter 36
Nematodes

A. Write true or false about the following.

1. *Amidostomum anseris* occurs in domestic and wild fowl and guinea fowls.

A. False. In duck and geese.

2. *Ascaridia galli* is the largest nematode of birds.

A. True.

3. Young birds are most seriously affected with gape worm.

A. True.

4. An occasional aesthetic problem with *Ascaridia galli* is the finding of this parasite inside of the hen's egg.

A. True.

5. *Amidostomum anseris* is quite host specific and attempts to infect a variety of other hosts were unsuccessful.

A. True.

6. Larvae of *Ascaridia galli* do not hatch but moult inside the eggs until they reach the L3 stage.

A. True.

7. *Ascaridia galli* is common for infection of older birds.

A. True.

8. *Heterakis gallinae* is also known as red worm because of its colour.

A. False. White colour and known as cecal worm.

9. Dietary deficiencies, such as vitamins A, B and B_{12}, minerals and proteins, predispose to ascaridiosis in poultry.

A. True

10. *Subulura brumpti* has direct life cycle.
A. False. indirect
11. *Epomidiostomum uncinatum* occurs in the intestine of domestic and wild ducks.
A. False. It occurs in gizzard.
12. The infective larvae of *Syngamous trachea* is very much resistant to desiccation.
A. False. Not resistant.
13. Tetramisole is the drug of choice for the treatment of eye worm infection in poultry.
A. True.
14. Grasshoppers and various beetles act as intermediate host in the life cycle of *Cheilospirura hamulosa*.
A. True.
15. *Dispharynx spiralis* generally occurs in the wall of the gizzard of the fowl, turkey, pigeon, guinea fowl etc.
A. False. It occurs in proventriculus and oesophagus.

B. Match the followings.

1.	Hair worm	a) *Ascaridia galli*
2.	Pin worm	b) *Cyathostoma bronchialis*
3.	Largest nematode of birds	c) *Cheilospirura hamulosa*
4.	Eye worm of poultry	d) *Tetrameres americana*
5.	Forked worm	e) *Heterakis gallinae*
6.	Cecal worm	f) *Oxyspirura mansoni*
7.	Sexual dimorphism	g) *Capillaria annulata*
8.	Thorny-headed worms	h) *Subulura brumpti*
9.	Gizzard Worms	g) *Polymorphus boschadis*
10.	Gape worm of duck	i) *Syngamus trachea*

Ans: 1. g 2. h 3. a 4. f 5. i 6. e 7. d 8. g 9. c 10. b

Fill in the blanks

1. The chief economic importance of the cecal worm (..................) of poultry lies in its role as a carrier of blackhead organism ....
2. No marked ill effects are seen in *Heterakis gallinae* unless in heavy infections in fowl and in other birds, but in pheasant another species of this genus, ... is very pathogenic causing .... typhlitis that may leads to diarrhoea, wasting, emaciation and death.
3. .... is characterized by parasitic generation consisting of only parthenogenic females in the intestine of the avian host.
4. Species of the genera *Ascaridia*, ... (cecal worm), *Capillaria* and ....(gape worm) are generally the most common .... parasites encountered in commercial poultry operations, sometimes with high prevalence rates.
5. An occasional aesthetic problem with .... is the finding of this parasite inside of the hen's egg.
6. .... is found in the caeca of chickens, turkeys and other birds. The worms are small, white and measure ¼ to ½ inch in length.
7. *Syngamus trachea* is a .... coloured round worm that attach to the trachea (windpipe) of birds and causes the disease referred to as ....
8. The gape worms can easily block the .... , so they are particularly harmful to .... birds with characteristic open-mouth breathing.
9. *Ascaridia galli* is a .... parasite that occurs in the .... of poultry.
10. The eggs of *A. galli* have thick, ... shells that are resistant to .... and persist for a long time in the environment.
11. Transport hosts such as .... are thought to play a role in transmission of *A. galli* and hence, ...

birds tend to have a higher risk of infection.

12. The infective eggs are ingested by a chicken, where it reaches the .... and hatches.
13. Temperature, ... levels and pH are thought to be triggering factors that signal the larva to hatch from its egg.
14. *Ascaridia galli* occurs in .... , whereas *Heterakis gallinae* occurs in .... of poultry.
15. The pin worm of poultry is .... That occurs in the ....of the fowl, turkey, guinea-fowl etc.
16. The buccal capsule of Syngamus trachea is ....shaped, bearing .... teeth at its base.
17. *Ornithostrongylus quadriradiatus* occurs in the ...., ..., and small intestine of the pigeon throughout the world telamon of its is .... shaped.
18. *Cyrnea spp.* óccur in the .... and .... of variety of birds and that parasites utilise the cockroaches as ....
19. *Oxyspirura mansoni,* commonly known as .... of poultry occurs under the ....of fowl, turkeys etc. and its pharynx resembles a .... in shape.
20. *Cheilospirura hamulosa* occurs in the .... of the fowl and turkey in most countries and its cuticle of anterior part is ornamented with ....
21. Soft, yellowish-red nodules in the thinner parts of the gizzard wall are most frequently seen in .... infection.

Ans. 1. *Heterakis gallinae, Histomonas meleagridis* 2. *Heterakis isolonche,* nodular 3. *Strngyloides avium* 4. *Heterakis, Syngamus,* nematode 5. *Ascaridia gall* 6. *Heterakis gallinae* 7. red, gapes 8. trachea, young 9. nematode, small intestine 10. albuminous, desiccation 11. earth-worms, free range 12. proventriculus 13. carbon dioxide 14. small intestine, caeca 15. *Subulura brumpti,* caeca 16. Cup, six to ten 17. crop, proventriculus, cross 18. Gizzard, proventriculus, intermediate host. 19. Eye worm, nictitating membrane, hourglass. 20. Gizzard, cordons. 21. *Cheilospirura hamulosa*

Chapter 37
Protozoa

A. Write true or false about the followings.

1. Coccidiosis is one of the most important disease of poultry belong to the genus Coccidia.

A. False. Genus *Eimeria.*

2. The life cycle of coccidian are direct, comprises a parasitic and a non parasitic phase.

A. True.

3. Sarcocysts are usually found in the skeletal muscles of birds.

A. True.

4. The sporulated oocyst of *Eimeria* have two sporocysts each containing four sporozoites.

A. False. Four sporocysts each containing two sporozoites.

5. Eimeria of birds and mammals have little host specificity.

A. False. Host specific.

6. Coccidiosis most frequently occurs in layers and breeders.

A. False. In young birds.

7. Chickens, turkeys, pheasants, Japanese quail and bobwhite quail all have their own species of coccidia.

A. True.

8. *Leucocytozoon* infects the blood and tissue cells of internal organs.

A. True.

9. Sarcosporidiosis can be a major problem in waterfowl.

A. True.

10. Avian plasmodia develop in mosquitoes mainly of the genera *Anopheles*.

A. False. *Culex*.

11. *Leucocytozoon* spp. are parasites of animals and birds.

A. False. Of birds only.

12. There are about 60 species of *Leucocytozoon* that are found in various birds.

A. True.

13. *Leucocytozoon* spp. are the most important intestinal protozoa of birds.

A. False. Blood protozoa.

14. *Leucocytozoon* spp. is not pathogenic in domestic and wild birds.

A. False. Pathogenic.

15. Control of the blackfly vector is most likely the best way to cut down *Leucocytozoon* infections.

A. True.

16. *Histomonas meleagridis* is classified as a flagellate but is one of a few protozoa that likewise have an amoeboid stage.

A. True.

17. In fact chickens are not, as a rule, highly susceptible to blackhead.

A. True.

18. The blackhead parasites can live for long periods in the cecal worm and its eggs.

A. True.

19. Tritrichomnas eberthi is a common trichomonad occurring in the caecum of the chicken.

A. True.

20. The chickens are experimentally infected with *Trypanosoma evansi*.

A. True.

21. Ionophorous antibiotics are coccidiostatic agents.

A. False. They are coccidiocidal.

22. Pin point to pin head sized grayish white foci are seen in *Eimeria necatrix* infection.

A. True.

23. *Wenyonella gallinae* infected chickens passed red coloured faeces.

A. False. Blackish green.

24. *Eimeria tenella* and *E. necatrix* are the less pathogenic coccidian parasite of domestic poultry.

A. False. Most pathogenic.

25. Amprolium is a coccidiostatic agent, whereas Clopidol is a coccidiocidal.

A. True.

26. Vitamins- Biotin, Riboflavin and Niacin restrict the severity of poultry coccidiosis.

A. False. They enhance the severity.

27. Vit. A, E & K restrict the severity of Coccidiosis.

A. True.

28. Coccivac is a live vaccine containing virulent strain of *Eimeria* spp.

A. True.

29. Toxoplasmosis in domestic poultry has not been reported.

A. False. Many reports.

30. Mortality rate of *Plasmodium gallinaceum* may reach up to 80%.

A. True.

31. The infective stage of *Haemoproteus columbae* is sporozoite.

A. True.

32. Adults pigeons are more susceptible to *Haemoproteus columbae* infection than nestlings.

A. False. Nestlings.

33. The definitive host of *Haemoproteus lophortyx* is duck and geese.

A. False. Quail.

34. Histomoniosis is a disease of young turkeys, whereas chicken is much more resistant.

A. True.

35. Coccidia of ducks belongs to genus *Eimeria* only.

A. False. Belong to *Eimeria, Wenyonella* and *Tyzzeria.*

36. Toxoplasma gondii is a highly host specific protozoa.

A. False. Least host specificity.

37. Cryptosporidium oocysts do not have sporocysts surrounding the sporozoites.

A. True.

38. Cryptosporidium oocysts do not sporulate within the host cell.

A. False. Sporulate within the host cell.

B. Fill in the blanks.

1. Coccidia are the .... parasites of the phylum ....

2. Most coccidian in poultry belong to the genus ....

3. The sporulated oocyst of *Eimeria* have ....sporocysts each containing ....sporozoites.

4. The most important protozoan parasites infecting poultry belong to the phylum ....which includes genera the *Eimeria,................Haemoproteus,* .... *Sarcocystis* and *Cryptosporidium.*

5. Avian coccidia of the genus ....are extremely ....specific.

6. Various dipteran intermediate hosts, such as simuliid flies and culicoid midges, are responsible for the transmission of ....infections in birds.

7. The three main species of *Leucocytozoon* are ....in ducks and geese, ...in chickens, and ....in turkeys although other

species are also described.

8. Black flies, ..., are important for the transmission of...................... to poultry such as turkeys and ducks.
9. Biting midges, ..., serve as intermediate hosts for *Haemoproteus nettionis* which infects domestic ducks.
10. Several genera of ....can transmit avian *Plasmodium* spp.
11. The pigeon fly (hippoboscid fly, louse fly), *Pseudolynchia canariensis,* is especially harmful to nestling pigeons and also transmits ...., which causes a malarial-like disease of pigeons.
12. This fly is unusual in that the ....matures inside the female and pupates immediately upon ejection.
13. This protozoan parasite.........................., the agent that causes blackhead apparently is carried in and is transmitted from bird to bird through the eggs of cecal worm.
14. Avian plasmodia develop in mosquitoes of the genera .... and .... , and rarely in ....
15. Rectal coccidiosis of poultry is caused by ....
16. Blackhead is a disease caused by a protozoan parasite, ...
17. Avian trichomoniosis of upper digestive tract caused by ....is a disease of ....
18. Avian trichomoniosis of lower digestive tract is caused by ....
19. *Histomonas meleagridis* is transmitted by embryonated eggs of ....
20. For controlling the drug resistance in poultry coccidiosis ....programme should be practiced.

21. The three stages of *Histomonas meleagridis* are .... , ... and ....

22. ....coloured droppings in turkey is found in *Histomonas meleagridis* infection.

23. *Hexamita meleagridis* has .... anterior flagella and two nuclei and this parasite multiplied by ....

24. ...., ...and .... are necessary for the sporulation of oocyst.

25. The most prevalent and damaging coccidian in geese are *Eimeria truncata,* which causes.....................coccidiosis and *Eimeria anseris,* which causes.................coccidiosis.

26. The most frequently occurring species of coccidia in pigeons is ....

27. The ....of *Haemoproteus columbae* are elongate and ....shaped that partially encircle the nucleus of RBC.

28. The schizogony of Leucocytozoon spp, occurs in the endothelial and parenchymatous cells of ...., ..., ... and other organs of avian hosts.

29. The common avian trypanosomes are ...., ... and ....

30. Turkey malaria is caused by ....and ....

Answer : 1. protozoan , Apicomplexa 2. *Eimeria* 3. four , two 4. Apicomplexa, *Leucocytozoon, Toxoplasma* 5. *Eimeria,* host 6. *Leucocytozoon* 7. *L. simondi, L. caulleryi, L. smithi* 8. *Simulium* spp, *Leucocytozoon* spp 9. Culicoides spp. 10. mosquitoes 11. *Haemoproteus columbae* 12. larvae 13. *Histomonas meleagridis* 14. *Culex, Aedes, Anopheles* 15. *Eimeria brunette* 16. *Histomonas meleagridis.* 17. *Trichomonas gallinae,* Pigeon nestlings. 18. *Tetratrichomonas gallinarum* 19. *Heterakis gallinarum.* 20. Shuttle 21. Invasive stage,

Vegetative Stage and resistance phase. 22. Sulpher yellow. 23. Six, Longitudinal binary fission. 24. Temperature, Oxygen and moisture 25. Renal , intestinal 26. *Eimeria labbeana* 27. Gametocytes, sausage 28. Liver, heart , kidney. 29. *Trypanosoma avium, T. calmetti* and *T. gallinarum 30. Plasmodium durae and P. griffithsi*

C. Multiple Choice Questions

1. *Histomonas maleagridis* is a (amoeboid/ciliate) protozoa .
2. Histomoniosis is mainly a disease of (Turkey/ chicken).
3. In *Hexamita meleagridis* infection the affected poults passed (Foamy & Watery/ red & semi liquid) faeces.
4. Avian trypanosomes are (highly/non) pathogenic.
5. *Tetratrichomonas gallinarum* affects the (lower/upper) digestive tract of birds.
6. *Trichomonas gallinae* is a disease of (adult/young) pigeons.
7. Sulphadimidine and sulphaquinoxaline have (coccidiocidal/coccidiostatic) action.
8. The most immunogenic species among poultry coccidia is (*Eimeria maxima/E. tenella*).
9. The rounded up sporozoite is known as (merozoite/ trophozoite).
10. The sporulated oocyst of *Wenyonella sp* contains (Two/ Four) sporocysts.

Answer : 1. Amoeboid, 2. Turkey 3. Foamy & Watery 4. Non 5. lower 6. young , 7. cocciodiostatic 8. *Eimeria maxima* 9. trophozoite 10. Four

D. Match the followings.

1.	Caecal coccidiosis	A. *Leucocytozoon simondi*
2.	Black head of turkey	B. *Haemoproteus columbae*
3.	Avian malaria	C. First generation schizont
4.	*Pseudolynchia canariensis*	D. *Eimeria tenella*
5.	Cryptozoites	E. *Histomonas meleagridis*

6.	Black flies	F. Second generation schizont
7.	Monensin	G. *Hexamita meleagridis*
8.	Metacryptozoites	H. *Cryptosporidium* spp.
9.	Autoinfection	I. Ionophorus compound
10.	Infectious catarrhal enteritis	J. *Plasmodium gallinaceum*

Ans. 1. D 2. F 3. K 4. B 5. C 6. A 7. J 8. G 9. I 10. H

E. Answer the following questions in brief.

1. In which way parasitic diseases are different from viral and bacterial diseases?

A. Parasitic diseases differ from viral and bacterial diseases by the following main characteristics features:

- Most of the parasites produce chronic/obscure infections to maintain their existence/survivality.
- Have complicated life cycles consisting different host and stages of parasites.
- Epidemiology and transmission are also different and diverse.
- Lack of useful serological diagnosis due to complex nature of antigens/more prominent immune evasion.
- Quarantine/disinfections are of little value.

2. What are the main criteria on which species of coccidian are identified?

A. Species of coccidian are generally identified on the basis of following main characters:

- sporulation time
- oocyst morphology
- host specificity
- immune specificity
- site and appearance of lesions
- prepatent period etc.

3. How the method of rearing influence in parasitism of birds?

A. The method of rearing has a marked influence on the particular parasitic problem. Confinement rearing tends to favour those parasites with short life cycles and direct transmission such as *Eimeria* spp., *Ascaridia galli, Heterakis gallinarum,* and *Capillaria* spp. The rearing of free-range birds or back yard flocks provides an opportunity for parasites (cestodes and trematodes) that have indirect life cycle. Birds can pick up infections through ingestion of intermediate host viz. flies, beetles, ants, snail and slugs etc which are in abundance in such environment. It should be remembered that parasitism in poultry is a flock problem; what happens to an individual bird is of relatively little economic importance.

4. Enumerate some of the important genera protozoan parasites of birds?

A. The most important protozoa infecting poultry belong to the phylum Apicomplexa which includes genera the *Eimeria, Leucocytozoon, Haemoproteus, Toxoplasma, Sarcocystis* and *Cryptosporidium.*

Flagellates such as *Histomonas, Trypanosoma, Trichomonas, Chilomastix* and *Hexamita* and amoeba of the genera *Entamoeba* and *Endolimax* are also found.

A microsporidian, *Encephalitozoon cunicule* has recently been reported from chickens.

5. Which parasitic disease you consider the most important in poultry and why?

A. Without question Coccidiosis is probably the most important parasites of poultry in terms of distribution, frequency and economic losses due to mortility as well as morbidity. All poultry production units, large or small, the ubiquitous presence of coccidia with the extreme reproductive ability of these intracellular parasites. Although, mortality from coccidiosis, now a days is under control by regular anticoccidial medications,however, besides the pathological effects, a huge economic losses are incurred world wide from reduced weight gain and feed efficiency and the cost of treatment.

6. Name three most important malaria parasites of domestic fowl and their vectors?

Over 65 species of *Plasmodium* have been described from different birds, less than 35 valid. *Plasmodium gallinaceum, P. juxtanucleare* and *P. durae* are the three most pathogenic malaria parasites for domestic fowl and may cause mortality of over 90%. Avian plasmodia develop in mosquitoes of the genera *Culex* and *Aedes,* and rarely in *Anopheles.*

7. Name some important nematodes in poultry industries?

Species of the genera *Ascaridia, Heterakis, Capillaria* and *Syngamus* (gape worm) are generally the most common nematodes encountered in commercial poultry operations, sometimes with high prevalence rates. There other genera such as *Dispharynx, Gongylonema, Tetrameres, Subulura* and *Trichostrongylus* are found far more frequently than in confinement facilities, especially those species that utilize an intermediate host.

8. What is the common fowl tick and its role?

The soft-bodied or fowl ticks of the genus *Argas* are the most important ticks in poultry, although many species of hard ticks will feed intermittently on poultry. Blood loss from the feeding of larvae and nymphs of soft ticks may even cause fatal anaemia. In many tropical countries, it has prevented the rearing of imported breeds. In addition, the fowl tick has been shown to transmit a number of spirochete, piroplasmosis, rickettsial and bacterial diseases of importance in many parts of the world.

9. What are the species of genus Leucocytozoon mostly found in birds?

Leucocytozoon smithi
Leucocytozoon simondi
Leucocytozoon caulleryi

L. smithi affects turkeys, *L. simondi* affects geese and ducks, and *L. caulleryi* affects chickens.

10. What are the synonyms of *Histomonas maleagridis* infection?

A. Histomoniasis, infectious enterohepatitis, black head disease.

11. Who is known as the "Father of Avian Coccidiosis Research"?

A. E.W. Tyzzer.

12. What are the developmental stages of Coccidian parasite?

A. Oocyst, sporocyst, sporozoites, schizont, merozoite, gametocyte.

13. Accumulation of a greenish fluid or cheesy materials in the mouth and crop and this may exude from the beak and there is a pendulous crop and foul odour from mouth. Give the name of the causative agent and its transmission to the birds.

A. *Trichomonas gallinae*, through crop's milk, , drinking water.

14. For excystation of oocyst of coccidia what are the stimuli required?

A. Two stimuli. One is provided by CO_2 and the second by trypsin and bile.

15. In which way Cryptosporidium differs from Eimeria sp. infecting poultry?

A. a. Intracellular stages are confined to the microvillus region of the host cell.

b. Oocysts sporulate within the host cell and are infective when released in the faeces.

c. Majority of Cryptosporidium oocysts are thick walled form unlike thin walled in *Eimeria sp.*

d. Auto infection occurs in *Cryptosporidia sp.*

e. Small number of oocysts may produce severe infection even the life threatening disease in immunidefficient host.

f. Mucosal epithelium of wide variety of organs may be affected.

Chapter 38
Fungal Diseases

A. Fill in the blanks.

1. .... is by far the most commonly encountered mycosis of birds and is an economically important .... disease of poultry.

A. Aspergillosis, respiratory

2. Dermatophytosis affects the .... and is the only .... and zoonotic mycosis.

A. integument, contagious

3. There are approximately .... subtypes of genus Aspergillus of which A. .... is the principal agent for causing aspergillosis in poultry.

A. 600, *fumigatus*

4. Spread of aspergillus is worldwide and should be considered as a .... , as it is found everywhere in nature due to its basic nutrient requirements.

A. commensal

5. While the optimum temperature for growth of aspergillus is .... °C, however it thrives anywhere from .... °C, and the higher the ...., the better it grows.

A. 40, 10-60, humidity

6. Aspergillosis in birds is usually confined to the lower .... system with florid lesions in the .... and ....

A. respiratory, air sacs, lungs

7. In young poultry, aspergillosis is referred to as .... pneumonia.

A. brooder

8. The fungi produce spores-conidia with a diameter of ....µm.

A. 2-3

9. In turkeys, we know that the .... produced by aspergillosis are toxic and immunosuppressive.

A. gliotoxins

10. .... aspergillosis occurs in adult birds particularly in ....

A. chronic, turkeys.

11. A. *fumigatus* produces a number of .... enzymes that can degrade host tissues, especially components of the .... matrix.

A. proteolytic, extracellular

12. Approximately 50% of turkey poults died following a .... minute aerosol exposure to .... of *Aspergillus fumigatus.*

A. 10, conidia

13. Candidiasis is the principal fungal infection of the .... tract of poultry.

A. digestive

14. Birds are particularly susceptible to oral and .... Candidiasis, which resembles ....in humans.

A. crop, thrush

15. ....is the primary agent of thrush, and comprise ....% of other species isolated.

A. *Candida albicans,* 95%

16. *Candida albicans* is a ....fungus readily isolated from the .... surfaces of .... digestive tract of birds.

A. commensal, mucosal, upper

17. Gross lesions of candidiasis are mostly confined to the ...., ... And ....

A. crop, proventriculus, gizzard

18. In candidiasis, the crop and proventriculus have whitish thickened areas that are often described as having a .... appearance.

A. turkish towel

19. ....and ....stains are helpful in visualizing the *Candida* morphology in tissue sections.

A. Periodic acid Schiff, Gomori's methanamine silver stain

20. *Candida* lesions occur most frequently in the ....and consist of thickening of ....with whitish, raised circular or ....formations.

A. crop, mucosa, rugose

21. Candida species can be identified by .... utilization panels.

A. substrate

22. The primary etiologic agent of favus is ....

A. *Microsporum gallinae*

23. The etiologic agent of favus was first described in 1881 as ...., later known as ...., then ...., and eventually accepted as ....

A. *Epidermophyton gallinae, Achorion gallinae, Trichophyton gallinae, Microsporum gallinae*

24. Care should be taken when handling birds with favus to prevent .... transmission.

A. zoonotic

25. ....is a sporadic fungal encephalitis of birds caused by the dematiaceous thermophilic fungus ....

A. Dactylariosis, *Dactylaria gallopava*

26. *Histoplasma capsulatum* is a .... fungus that

occurs as an environmental .... and as a .... in its warm blooded host.

A. dimorphic, mould, yeast

27. Histoplasmosis infection is acquired by ....of conidia produced by the ....form.

A. inhalation, mould

28. Ochratoxin is a .... derivative linked to............................

A. dihydroisocoumarin, L-â-phenylalanine

23. The name of ochratoxin derives from ...., the first fungus shown to produce it.

A. *Aspergillus ochraceus*

30. Ochratoxin A inhibits .... synthesis, produces acute proximal tubular epithelial necrosis in the kidneys, and inhibits normal renal .... secretion.

A. protein, uric acid

B. Multiple choice

1. The principal agent of aspergillosis in poultry is (*Aspergillus fumigatus / A. flavus*).

A. *Aspergillus fumigatus*

2. Aspergillosis is a (transmissible / not transmissible) disease.

A. not transmissible

3. Manifestations of aspergillosis depend on (species or strain / organ or system involved).

A. organ or system involved.

4. Cutaneous lesions as a manifestation aspergillosis are (common / rare) in avian species.

A. rare

5. (Pulmonary / systemic) aspergillosis is the most common form of the disease in avian species.

A. Pulmonary

6. Egg embryo are quite (susceptible / resistant) to infection by *Aspergillus fumigatus* during incubation.

A. susceptible

7. The conidia of (*Aspergillus fumigatus* / *A. flavus*) are about 2-3µm in diameter.

A. *Aspergillus fumigatus*

8. The size of *Aspergillus fumigatus* conidia is (greater / smaller) than that of *A. flavus*.

A. smaller

9. In a mycological survey of crops, *Candida albicans* comprised (25%, 75%, 95%) of all other *Candida* species.

A. 95%

10. Candidiasis is the principal fungal infection of the (upper / lower) digestive tract of poultry.

A. upper

11. Candidiasis is also referred to as (favus / thrush).

A. thrush

12. *Candida* cells stain (Gram positive / Gram negative).

A. Gram positive

13. Thrush tend to show (increased / decreased) feed consumption and (increased / decreased) water consumption.

A. decreased, increased

14. *Candida* lesions occur most frequently in the (crop / intestine).

A. crop

15. A raised circular (whitish / grayish) lesions in the crop are characteristic in canddiasis.

A. whitish

16. Favus is the term applied to the fungal infections of (digestive tract / respiratory tract / skin).

A. skin

17. The fungus belongs to the (perfect / imperfect) yeast group under the name of *Cryptococcus neoformans*.

A. imperfect

18. Young poultry are more sensitive to aflatoxin than are adult.

A.

C. Match the following :

A.

1.	Torulosis	A. Candidiasis
2.	Zoonotic mycosis	B. Aspergillosis
3.	Sour crop	C. Favus
4.	Brooder pneumonia	D. Mucor, Rhizopus
5.	Zygomycosis	E. Cryptococcosis

Ans. 1. E 2. C 3. A 4. B 5. D

B.

1.	Rugose formation in crop	A. *Microsporum gallinae*
2.	Dimorphic fungus	B. *Dactylaria gallopova*
3.	Skin scrapings	C. *Aspergillus flavus*
4.	Thermophilic fungus	D. *Histoplasma capsulatum*
5.	Gliotoxin	F. *Candida albicans*

Ans. 1. E 2. D 3. A 4. B 5. C

D. Write true or false about the followings.

1. Mycotic diseases in poultry are generally contagious in nature.

A. False. Not contagious except dermatophytosis.

2. Aspergillosis is a contagious and zoonotic disease.

A. False. It is not contagious or zoonotic.

3. As a saprophyte, aspergillus is not specifically a poultry pathogen and therefore grows everywhere that conditions are favourable.

A. True.

4. Acute aspergillosis is characterized by severe outbreaks in older birds.

A. False. In young birds.

5. Crossbred or outbred strains are more susceptible than inbred strains during an outbreak of aspergillosis in hatchery chicks.

A. False. Inbred strains are more susceptible.

6. Aspergillosis can be acquired in ovo.

A. True.

7. Infection of eye, brain, skin, joints and viscera are less common manifestation in avian aspergillosis.

A. True.

8. For aspergillosis, vaccines are not a practical alternative and none are commercially available.

A. True.

9. The aflatoxins include four closely related metabolites of *Aspergillus flavus* known as B_1, B_2, B_3 and B_4.

A. False. B_1, B_2, G and G_2.

10. The B_1 toxin is the most toxic and is of greatest concern to the poultry industry.

A. True.

11. Candida species have a world wide distribution and are the part of the microflora of the healthy digestive system of human, animals and birds.

A. True.

12. The term thrush is applied to *Aspergillus* infections.

A. False. It is *candida* infections.

13. The occurrence of avian candidiasis is sporadic but outbreaks can be costly.

A. True.

14. Candidiasis is sometimes referred to as moniliasis.

A. True.

15. The morphology of *Candida* in tissues is fairly characteristic in that mycelial and yeast forms are all present in lesion.

A. True.

16. Formation of germ tube which was observed in all the isolates of *Candida* is confirmatory for this fungus.

A. True.

17. A history of long-term antimicrobial therapy is supportive when *Candida* is suspected.

A. True.

18. Histoplasmosis is a contagious mycotic disease of animals and man.

A. False. It is not contagious.

19. The term favus usually is used to denote the disease in poultry.

A. True.

20. Favus is not a contagious disease.

A. False. It is contagious.

21. *Microsporum gallinae* can produce ringworm lesions in their human handlers.

A. True.

22. *Candida albicans* is a normal inhabitant of the human alimentary canal, as well as that of many species of lower animals.

A. True.

23. Cryptococcosis is a disease humans and animals.

A. True.

24. Prognosis is grave in case of cryptococcal meningoencephalitis.

A. True.

25. Commonly used synonyms of zygomycosis is dermatophytosis.

A. False. It is phycomycosis and mucormycosis.

26. Young poultry are less sensitive to aflatoxin than are adult.

A. False. Young are more sensitive.

27. Quail and turkeys are more sensitive to ochratoxicosis than chickens.

A. True.

28. Based on its acute toxicity ochratoxicosis is less lethal than aflatoxicosis.
A. False. More lethal.
29. Among the ochratoxins, ochratoxin A is the least toxic and produced in lesser quantities.
A. False. Most toxic and produced in greater quantities.
30. *Aspergillus flavus,* a fungi that cause various diseases of the skin and hair, referred to as favus.
A. False. It is *Microsporum gallinae.*

E. Answer the followings questions in brief.

1. Define aspergillosis.
A. Aspergillosis is defined as a fungal disease caused by infection with the genus *Aspergillus,* which is composed of about 600 species.
2. What are the synonyms used for aspergillosis?
A. In young poultry, the disease is commonly called as "brooder pneumonia". Other names include fungal or mycotic pneumonia, pneumonomycosis, bronchomycosis, asper and air sac etc.
3. What is the principal agent causing aspergillosis in poultry?
A. *Aspergillus fumigatus.*
4. Name few other species besides *Aspergillus fumigatus* which are frequently causing aspergillosis in birds.
A. *Aspergillus flavus, A. niger, A. glaucus, A. nidulans, A. terreus, A. nigrescens, A. amstelodami etc.*
5. What is the most common form/ organs affected in aspergillosis seen in avian species?
A. Pulmonary aspergillosis is the most common form affecting lower respiratory system with florid lesions in air sacs and lungs.
6. What are the other forms/organs affected in aspergillosis may also be encountered?

A. Other than pulmonary form, systemic, dermatitis, osteomycosis, opthalmitis and encephalitic aspergillosis may also be encountered.

7. What are the culture media used for *Aspergillus?*

A. The fungus grows quite readily on Sabouraud's dextrose, Czapek's solution agar or potato dextrose agar culture media at room temperature, at 37°C, and higher.

8. Write down the colony characters of *Aspergillus*?

A. The colonies are green to bluish green at first and darken with age so as to appear almost black. The colonies vary from velvety to floccose. The conidiophores are short, up to 300μ long by 2-8μ in diameter, the vesicles are apical flask-shaped up to 20-30μ in diameter, the sterigmata are 6-8 by 2-3μ in diameter, in chains forming solid columns up to 400 by 50μ.

9. What are the main two forms aspergillosis in poultry?

A. Aspergillosis is encountered in poultry in two main forms. Acute outbreaks in which there are high morbidity and high mortality may occur particularly in young birds. In chronic form, adults especially, an occasional bird in a flock or aviary may become affected.

10. Define candidiasis.

A. Candidiasis is a mycotic disease of the upper digestive tract of birds caused by *Candida* genus, principally *C. albicans.* It is characterised by thickening and white plaques on the mucosa, especially in the crop but sometimes in the proventriculus and gizzard.

11. What are the synonyms of *Candida* infections?

A. Moniliasis, candidiasis, thrush, sour crop, crop mycosis, oidiomycosis, stomatitis oidica, muguet (France), soor (Germany) etc.

12. What are the gross lesions of crop mycosis?

A. Lesions are generally confined to the upper areas of the digestive tract, mouth, esophagus, and, primarily, the crop, may have grayish-white, loosely attached, plaque-like areas on their internal surfaces. Circular, raised, ulcerative nodules that appear as rose-like clusters may be within the crop, and the crop surface is often so unevenly thickened that it appears like texture of a Turkish bath towel or curds. Other areas of the upper digestive tract develop false membranes that resemble those which develop during diphtheria, areas of dead tissue, and contain considerable tissue debris.

13. What are the most commonly seen mycotoxicosis in commercial poultry?

A. Aflatoxicosis, ochratoxicosis and trichthecene mycotoxicosis.

Chapter 39

Chlamydiosis

A. Write true or false about the followings.

1. *Chlamydophila psittaci* is excreted in the feces and nasal discharges of infected birds.

A. True.

2. Some infected birds can appear healthy and shed the organism intermittently.

A. True.

3. Psittacosis typically causes influenza-like symptoms and can lead to severe pneumonia and nonrespiratory health problems.

A. True.

4. Patients who develop psittacosis require treatment with specific drugs.

A. True.

5. The disease psittacosis is very responsive to penicillin but is resistant to tetracycline.

A. False. Responsive to tetracycline.

6. This disease can be transmitted from infected birds to humans.

A. True.

7. Parrot disease, ornithosis, and chlamydiosis are other names for psittacosis.

A. True.

8. The infection can not be treated with antibiotics.

A. False. Treated with antibiotics.

9. *C. psittaci* in birds is often systemic and infections can be

inapparent, severe, acute or chronic with intermittent shedding.

A. True.

10. Psittacosis is primarily a lung disease but it can involve several organs. Some reports show that inflammation of the liver, lining of the heart cavity, the heart muscle, and the brain can occur.

A. True.

11. There are number of commercial vaccines available against psittacosis.

A. False. Not available.

B. Fill in the blanks.

1. *Chlamydophila psittaci* is a member of the family ....

A. *Chlamydiaceae.*

2. .... is also known as parrot disease, parrot fever, and ornithosis is a .... infectious disease caused by a called *Chlamydophila psittaci* which was formerly known as ....

A. Psittacosis, zoonotic, *Chlamydia psittaci.*

3. Currently there are .... serovars and .... genotypes described which in the future may prove of importance in the epidemiology of the disease in animals and humans.

A. 8, 9.

4. Psittacosis typically causes .... like symptoms and can lead to severe .... and nonrespiratory health problems.

A. Influenza, pneumonia.

5. The word psittacosis is derived from the Greek work Psittakos, which means ....

A. parrot.

6. Psittacosis is primarily a disease of .... but it

can involve several organs.

A. lung.

7. .... probably is the best alternative antibotic for persons for whom tetracycline is contraindicated especially in children aged less than 9 years and pregnant women.

A. Erythromycin.

8. Symptoms in birds include .... eyes, difficulty in breathing, ... droppings and .... urates.

A. inflamed, watery, green

9. Infection of psittacosis doesn't confer ...., which means a person regularly exposed to infected birds can become sick again.

A. immunity.

10. *C. psittaci* in birds is often .... and infections can be inapparent, severe, acute or chronic with .... shedding.

A. systemic, intermittent.

C. Answer the followings.

1. What are the synonyms of avian chlamydiosis?

A. It is called psittacosis if it infects psittacines (parrots, parakeets, cockatoos, etc.), the same malady is termed ornithosis if it infects other types of birds. Other names include parrot fever and chlamydiosis.

2. How chlamydia differ from other bacteria?

A. Chlamydiae differ from other kinds of bacteria in two important respects: they are considerably smaller and they behave more like viruses than bacteria. Chlamydiae can't exist outside a host cell. Once they have gained access to a suitable host cell, the bacteria 'reprogram' the cell's DNA to produce more Chlamydiae, which is how viruses multiply. This "microvarmit" is like a bacteria (in that it responds to antibiotics) as well as like a virus (in that it will grow only on a tissue culture, not on bacterial media).

3. What are the strains of *C. psittaci* prevalent in the world?

A. *C. psittaci* strains are similar in virulence, grow readily in cell culture, have 16S-rRNA genes that differ by <0.8%, and belong to eight known serovars. All should be considered to be readily transmissible to humans. *C. psittaci* serovar A is endemic among psittacine birds and has caused sporadic zoonotic disease in humans, other mammals and tortoises. Serovar B is endemic among pigeons, and has also been identified as the cause of abortion in a dairy herd. Serovars C and D are occupational hazards for slaughterhouse workers and for people in contact with birds. Serovar E isolates (known as Cal-10, MP or MN) have been obtained from a variety of avian hosts worldwide and although, they were associated with the 1920s–1930s outbreak in humans, a specific reservoir for serovar E has not been identified. The M56 and WC serovars were isolated during outbreaks in mammals.

4. How do birds pass on the infection to humans?

A. Human can become infected with *Chlamydia psittaci* by breathing in the organism when the urine, respiratory secretion, or dried feces of infected birds is aerosolized (i.e., dispersed in the air as very fine droplets or dust particles). Other sources of exposure include mouth-to-beak contact, a bite from an infected bird, and handling the plumage and tissues of infected birds.

5. Can humans transmit the infection to other humans?

A. Person-to-person transmission of the disease is rare. It may occur when a person is exposed to infectious, aerosolized droplets from another person experiencing paroxysmal (sudden, very forceful) coughing during the acute illness.

6. What are the clinical signs of psittacosis?

A. The usual duration between exposure to *C psittaci* and onset of illness ranges from 3 days to several weeks. However, active disease can appear with no identifiable exposure. Whether the bird has acute or chronic signs of illness or dies depend on the species of bird, virulence of the strain, infectious dose, stress factors, age, and extent of treatment or prophylaxis.

Signs of avian chlamydiosis are non-specific and include lethargy, anorexia, and ruffled feathers. Other signs include serous or mucopurulent ocular or nasal discharge, diarrhoea, and excretion of green to yellow-green urates. Severely affected birds may become anorectic and produce sparse, dark green droppings, followed by emaciation, dehydration, and death.

C. psittaci in birds is often systemic and infections can be inapparent, severe, acute or chronic with intermittent shedding. Symptoms in birds include "inflamed eyes, difficulty in breathing, watery droppings and green urates.

7. What are the occupations at risk for Psittacosis?

A. Psittacosis is an occupational health hazard for many people whose work brings them into contact with birds. Bird owners, pet shop employees, and veterinarians are at risk of the infection. Some outbreaks of psittacosis in poultry processing plants have been reported. These include: bird fanciers, pigeon fanciers, poultry production workers, poultry processing worker's, pet shop employees, quarantine facilities employees, veterinary clinics employees, diagnostic laboratories employees, racing pigeons keepers, public health inspectors, exotic and domestic bird breeders, bird dealers.

8. How is psittacosis recognized and treated?

A. For accurate diagnosis of psittacosis, a doctor must know that the person has been exposed to birds and that the suspected birds are infected with *Chlamydia psittaci*. Laboratory examinations can identify the organism and detect the signs of infection. Patients who develop psittacosis require treatment with specific drugs. The disease is very responsive to tetracycline but is resistant to penicillin.

9. How will you define a confirm case of avian chlamydiosis?

A. A confirmed case of avian chlamydiosis can be defined on the basis of one of the following:

1) Isolation of *C psittaci* from a clinical specimen.
2) Identification of chlamydial antigen by use of immunofluorescence (fluorescent antibody) in the tissues.
3) Increase in serologic titer (= 4-fold) in two specimens from the bird obtained at least two weeks apart and assayed simultaneously at the same laboratory, or
4) Identification of *Chlamydiaceae* within macrophages in smears or tissues stained with Gimenez or Macchiavello stain.

10. What are the preventive measures of avian chlamydial infection?

A. Once psittacosis has been diagnosed, it is important to treat the birds and their environment. Remember that healthy birds may be harbouring the bacteria as well. These include:

- Avoid unnecessary handling of sick birds.
- Avoid breathing in any dust from dried bird droppings, feathers or cage dust.
- Isolate sick birds from the rest of the flock.
- Treat infected birds with appropriate antibiotics for at least one month.
- Clean cages with appropriate disinfectants, since the bacteria can live for several months in shed feathers and droppings.
- Wear masks and gloves while cleaning the cages to prevent infection.
- Clean the cages regularly, using plenty of water to minimise the risk of floating danger. Always wash hands thoroughly after tending to birds.

11. What can be done to prevent the spread of psittacosis?

A. If birds are kept as pets, clean the cage often so that fecal material does not accumulate, dry up and become airborne. Current laws require that members of the parrot family that are imported from foreign countries be kept in a bird quarantine station prior to sale. During the quarantine, they are given feed containing tetracycline to reduce the risk of infection, but the duration of treatment is generally shorter than the 45 days necessary to completely treat the bird. Therefore, federal health authorities have recently recommended that breeders and importers ensure that all domestic nestlings and imported birds receive the special tetracycline feed for 45 continuous days in order to prevent the spread of psittacosis from birds to humans.

Chapter **40**

Spirochetes

A. Fill in the blanks.

1. *Borrelia anserina* is a helical filiform .... bacteria.

A. spirochete.

2. ....causes a tick-borne, nonrelapsing spirochetosis in avian species.

A. *Borrelia anserina.*

3. *Borrelia anserina* is a highly motile, ... bacterium.

A. helical.

4. *Borrelia anserina* is a helical bacteria with about .... spirals.

A. 5-8.

5. Birds infected with virulent strains of *B. anserina* are visibly sick, with .... of comb and wattles.

A. cyanosis.

6. During the initial stages of the disease, affected birds pass fluid or .... droppings with excess .... and ....

A. green, bile, urates.

7. Marked enlargement and mottling of the .... is the most characteristic lesion of borreliosis in birds.

A. spleen

8. In Microbiological laboratories, *B. anserina* bacteria can be grown on special protein enriched mediums which are rich in ovoalbumins or animal tissue that countains .... , in .... conditions.

A. Myoglobin, anaerobic

9. *Borrelia anserina,* which causes .... in chickens and other birds, is propagated by .... of the genus ....

A. spirochetosis, ticks, *Argas.*

10. Intestinal spirochaetes of poultry are .... gram................, ... shaped bacteria.

A. anaerobic, negative, helix

11. Recently, pathogenic intestinal spirochetes of birds previously designated as the genus .... have been moved into the family................. , genus..........................

A. *Serpulina,* Brachyspiraceae, *Brachyspira*

12. Use of an .... system and numerical .... coding system has been used in differentiating intestinal spirochetes of avian and mammalian isolates.

A. API-ZYM, 5- digit.

B. Write True or False on the follwing statements.

1. Attempts to transmit *Borrelia anserina* with *Amblyomma* ticks were successful.

A. False. It was unsuccessful.

2. *Borrelia anserina* is a non motile, helical bacterium.

A. False. It is highly motile.

3. *Borrelia anserina* is a filiform bacteria.

A. True.

4. *B. anserina* can be coloured with the May Grunwald-Giemsa method.

A. True.

5. *Borrelia anserina* a bacterial species that causes spirochetosis of fowls; found in the intestine of infected geese, ducks, other fowl.

A. False. It is found in the blood.

6. *Borrelia anserina* can not penetrate unbroken skin.

A. False. Virulent strains can penetrate.

7. In fowl spirochetosis, recovered birds are usually serve as carrier of the disease.

A. False. Organisms disappear from circulations and tissues.

8. Borreliosis in avian species is a mosquito borne disease.

A. False. Tick borne.

9. Young birds are affected more severely than older ones.

A. True.

10. The most common vector is *Argas persicus*, the common fowl tick, but other *Argas* spp transmit the disease in different geographic areas.

A. True.

11. All borreliae are arthropod-borne.

A. True

12. Avian intestinal spirochetosis (AIS) is an acute septicaemic disease characterised by spiral bacteria.

A. False. Subacute to chronic nonsepticaemic disease.

13. The disease is caused by a homogeneous group of spirochaetes in the small intestine.

A. False. Heterogeneous group of spirochaetes in the rectum and/or colon.

14. Mechanism of pathogenicity for intestinal spirochetes are now completely understood.

A. False. Incompletely.

C. Answer the followings.

1. Define borreliosis (avian spirochaetosis).

A. Avian spirochetosis is an acute, febrile, septicaemic, nonrelapsing, tick borne bacterial disease that affects a wide variety of birds caused by *Borrelia anserina*.

2. What are the synonyms of *Borrelia anserina* infection?

A. Bird borreliosis or spirochetosis.

3. What are the common *Borrelia* found in different hosts?

A. *B. anserina* – causes fowl spirochetosis.

B. burgdorferi – causes lyme disease in humans and animals.

B. recurrentis – causes relapsing fever in humans.

B. theileri – cause of theileriasis in cattle.

3. What are the characteristic clinical signs/lesions of fowl spirochetosis?

A. Clinical signs varies depending on the strain virulence of the spirochete. Birds infected with virulent strains are visibly sick. Signs include listlessness, depression, somnolence, moderate to marked shivering, increased thirst and cyanosis of comb and wattles. Young birds are affected more severely than older ones. During the initial stages of the disease, there is usually a green or yellow diarrhoea containing excess bile and urates. Later, birds may develop paresis or paralysis, become anaemic, somnolent to comatose. The course of the disease is 1-2 wk. In tick-infested geographic areas, morbidity can approach 100% and mortality may be 33-77%. Egg production may be reduced by 5-10%, with a higher number of small eggs. Marked enlargement and mottling of the spleen is the most characteristic lesion.

4. What are the most effective antibiotics for treatment of fowl spirochetosis?

A. Tetracyclines, penicillins and doxycycline are the most effective antibiotics for treatment of spirochetes.

5. Define Avian intestinal spirochetosis (AIS).

A. Avian intestinal spirochetosis (AIS) is a subacute to chronic, nonsepticaemic intestinal disease characterised by spiral bacteria.The disease is caused by a heterogeneous group of spirochaetes in the rectum and/or colon.

6. Name some common species of avian intestinal spirochetes.

A. *Brachyspira hyodyseneriae, B.intermedia, B. pilosicoli, B.alvinipulli, B. innocens, B. murdochii.*

7. What are the clinical signs Avian intestinal spirochaetes?

A. Avian intestinal spirochaetes can be divided into three pathotypes, causing: subclinical infection; mild to moderate or severe clinical disease. In layers, wet faeces, diarrhoea, pasted vents, retarded growth, delayed onset of lay, production of dirty faecal-stained eggshells, reduced egg weight and reduced egg carotenoid content may all be evident. In turkeys, broilers and broiler breeders are similar to those in layers except that growth retardation is more severe.

8. What are the diagnosis approaches for AIS?

A. Spirochaetes can be normal flora or produce subclinical infections, characteristic clinical signs and lesions must be present in conjunction with visual demonstration of helical bacteria by dark-field or light microscopy for a presumptive diagnosis of AIS. Confirmation of bacteria as spirochaetes should be through visualisation of distinctive ultra-structural features, demonstration of spirochaete antigens or by isolation in culture. Culture from fresh faecal droppings or caecal mucosa followed by further characterisation is necessary to confirm a diagnosis. In poultry, spirochaetes identified from faecal specimens should be distinguished from other spiral bacteria such as Campylobacter, Arcobacter, Helicobacter and Spirillum. In case of chronic diarrhoea or pasted vents, nutritional problems such as excess dietary salt, fats or soya bean meal should be investigated. Other causes of chronic diarrhoea include enteric salmonellosis, colibaccillosis and coccidiosis.